The
OXFORD
Children's Encyclopedia of
Plants &
Animals

OXFORD

UNIVERSITY PRESS

OXFORD
UNIVERSITY PRESS

Great Clarendon Street, Oxford OX2 6DP

Oxford University Press is a department of the University of Oxford.
It furthers the University's objective of excellence in research, scholarship,
and education by publishing worldwide in

Oxford New York

Athens Auckland Bangkok Bogotá Buenos Aires Calcutta
Cape Town Chennai Dar es Salaam Delhi Florence Hong Kong Istanbul
Karachi Kuala Lumpur Madrid Melbourne Mexico City Mumbai
Nairobi Paris São Paulo Singapore Taipei Tokyo Toronto Warsaw

with associated companies in Berlin Ibadan

Oxford is a registered trade mark of Oxford University Press
in the UK and in certain other countries

British Library Cataloguing in Publication Data

Data available

ISBN 0-19-910607-X

1 3 5 7 9 10 8 6 4 2

Typeset by Oxford Designers and Illustrators
Typeset in Photina and Rotis
Printed in Britain by Butler and Tanner

Contents

Contributors

Editor
Ben Dupré

Coordinating editors
Ian Crofton
Joanna Harris

Proofreaders
Helen Maxey
Susan Mushin

Indexer
Ann Barrett

Design
Jo Cameron
Oxford Designers and Illustrators

Art editor
Hilary Wright

Assistant art editor
Jo Samways

Cover design
Jo Cameron

Photographic research
Charlotte Lippmann

Consultants
Bridget Ardley
John R. Brown
Esmond Harris
Sonia Hinton
Peter Holden
Dr George McGavin
Dr Stuart Milligan
Colin Mills
Dr Chris Norris
Joyce Pope
Elspeth Scott
Michael Scott
Bob Unwin

Authors
Bridget Ardley
Jill Bailey
Professor Chris Baines
Henry Bennet-Clark
Dr Michael J. Benton
Judith Court
Ian Crofton
David Glover
Sonia Hinton
Peter Holden
Dr Nick Middleton
Joyce Pope
Judy Ridgway
Theodore Rowland-Entwistle
Michael Scott
Andrew Solway
Dr Philip Whitfield

Acknowledgments

Key t top; b bottom; c centre; r right; l left
NHPA = Natural History Photo Agency; SPL = Science Photo Library; OSF = Oxford Scientific Films

Photos are reproduced by kind permission of:
Front cover Tony Stone, Tim Davis. 7t NHPA, M. I. Walker. 8t NHPA, Daniel Heuclin. 9t NHPA, Steve Robinson. 9bl NHPA, G. I. Bernard. 12t NHPA, Steve Robinson. 15t NHPA, Stephen Dalton. 15b NHPA, John Shaw. 17t Images of Africa, David Keith Jones. 17b SPL, Food and Drug Administration. 19t NHPA, Alan Williams. 19b NHPA, Orion Press. 20t NHPA, L. Hugh Newman. 21t SPL, Ken Eward. 22t SPL, CNRI. 24t NIIPA, John Shaw. 25b Image Bank, Margaret Mead. 26t NHPA, Stephen Krasemann. 27t NHPA, Gerard Lacz. 28b BBC Natural History Unit, Dietmar Nill. 31t NHPA, Kevin Schafer. 32t SPL, NASA. 32b NHPA, David Woodfall. 33t NHPA, Jeff Goodman. 34t OSF, Harold Taylor. 34b NHPA, B. Jones & M. Shimlock. 36t NHPA, Anthony Bannister. 38b NHPA, Daniel Heuclin. 40t OSF, David C. Fritts. 40b NHPA, Michael Leach. 41t Images colour library, 43r NHPA, Mirko Stelzner. 44b NHPA, Norbet Wu. 45c NHPA, E. A. Janes. 46tr Images of India. 46bl NHPA, Anthony Bannister. 47b Bridgeman Art Library, Downe House Kent. 48bl NHPA, Daniel Heuclin. 48br NHPA, Stephen Dalton. 49b NHPA, Norbet Wu. 53t NHPA, Stephen Dalton. p53c NHPA, John Shaw. 53b Heather Angel. 54t NHPA, Daniel Heuclin. 56t NHPA, Rod Plank. 57b NHPA, Stephen Dalton. 59t Sally and Richard Greenhill. 60b SPL, Philippe Plailly/Eurelios. 61t NHPA, Nigel Dennis. 62t NHPA, Laurie Cambell. 62b NHPA, Stephen Dalton. p63t NHPA, John Shaw. p63b NHPA, Christophe Ratier. 66b NHPA, Hellio & Van Ingen. 67b Woodfall Wild Images, Steve Austin. 68t Images of Africa, Johann Van Tonder. 68b Woodfall Wild Images, Nigel Hicks. 69t NHPA, Manfred Danegger. 69b NHPA, Stephen Dalton. 70t, 71b NHPA, Anthony Bannister. 72tr Robert Harding, N. A. Callow. 73t NHPA, B. Jones & M. Shimlock. 76t NHPA, E. A. Janes. 77cl NHPA, A. N. T. 80t NHPA, B. Jones & M. Shimlock. 80b NHPA, Daniel Heuclin. 82t NHPA, David Woodfall. 86l NHPA, Norbet Wu. 86tr Planet Earth, James D. Watts. 86b Woodfall Wild Images, Heinrich van Denberg. 87t NHPA, Anthony Bannister. 88t NHPA, Karl Switak. 89t NHPA, John Shaw. 89b NHPA, David Middleton. 90t NHPA, G. I. Bernard. 91t SPL, NOAA. 92t SPL, John Reader. 93l OSF, Douglas Faulkner. 93r Diane Fress, FRS. 95t National Geographic Image Collection, Stanley Breeden. 96t NHPA, Silvestris Fotoservice. 96b NHPA, T. Kitchin & V. Hurst. 97b NHPA, Kevin Schafer. 98b NHPA, Bryan and Cherry Alexander. 100cr NHPA, Melvin Grey. 102br SPL, Dr Jeremy Burgess. 103 SPL, Don Fawcett. 104t NHPA, Bruce Beehler. 104b NHPA, David E. Myers. 106t Planet Earth, David Maitland. 110b Woodfall Wild Images, Paul Kay. 112t NHPA, Silvestris Fotoservice. 112b Robert Harding, Minden Pictures. 114t, b Heather Angel. 115b NHPA, David Woodfall. 116t NHPA, Rich Kirchner. 117t SPL, Omikron. 119 NHPA, John Shaw. 120l NHPA, Stephen Krasemann. 120r NHPA, Dr Ivan Polunin. 123b NHPA, Christophe Ratier. 124t NHPA, Daniel Heuclin. 124b Vardon Attractions.

The illustrations are by:
Allen, Graham: 91b
Allington, Sophie: 109
Arlott, Norman: 19cl, 20b, 117b,
Baker, Julian: 35t, 77r, 95b
Barber, John: 56, 79b
Beckett, Brian: 16t, 33l
Beckett, Brian/Gecko: 72
Courtney, Michael: 64br
Gaffney, Michael: 99, 100t
Gecko Ltd: 14b, 29
Hinks, Gary: 108b
Hiscock, Karen: 58
Hutchins, Ray: 41b
Kennard, Frank: 37b, 73b, 75b, 83b, 106b
Linden Artists: 52
Loates, Mick: 49, 50
Madison, Kevin: 57
Milne, Sean: 13br, tr, 14t, 28t, 36b, 43tr, 52l, 54, 55, 58, 62ct, 64bl, 66t, 87b, 92, 93b, 100b, 101, 102, 105b
Milne, Sean/Richardson, Paul/Roberts, Steve: p122
Milne, Sean/Sneddon, James: 44t
Moore, David: 115tr
Ovenden, Denys: 38br
Pearson, Olive: all maps
Richardson, Paul: 11 cheetah, 13bl, tl, 25t, 26, 27, 35b, 37t, 42t, 45c, 67t, 71l, 81, 98t, 105t, 121
Roberts, Steve: 11, 12b, 16b, 18cl, 45b, 51, 75t, 76b, 78tr, cl, br, 79c, c, 83t, 84t, 108tr, 110, 111, 113t, 116b, 118, 121
Robinson, Andrew: 31b, 93tr
Sanders, Michael: 18tr, 42b, 82b, 84b, 104tr, 107t, 108c
Seymore, Steve: 60tl
Sneddon, James: 10 background, 59b, 60tr, 74, 90 background
Verrinder, Halli: 48
Visscher, Peter: 7b, 10, 24b, 74, 85r, 90 foreground
Weston, Steve: 21, 22, 23, 61b, 65, 70b
Wiley, Terry: 39, 47t, 94
Woods, Michael: 8b, 30b, 46br, 78tl, 113b
Woods, Michael/Gecko: 97t, 123t
Woods, Michael/Sneddon, James : 88b

Finding your way around

The *Oxford Children's Encyclopedia of Plants and Animals* has many useful features that will help you find the information you need quickly and easily.

The articles in the encyclopedia are arranged in alphabetical order from Algae to Zoos. When you want to find out about a particular topic, the first step is to see whether there is an **article** on it in the A–Z sequence. If there is no article, there are two things you can do.

First of all you can look at the **footers** at the bottom of the page.

These may include the topic you want, and give you the name of the article where you can find out about it. If there is no footer, the next thing to do is to look the topic up in the alphabetical **index** at the back of the book. This will tell you which page or pages you can look at to find out what you want to know.

The header tells you what articles are on the page, for quick reference.

Articles are arranged alphabetically, so that they are easy to find.

The **opening paragraph** gives a friendly introduction to the topic.

The **main text** gives an account of the topic in a continuous and readable way. Key terms are picked out in *italic text*.

Hippopotamuses

There are two kinds of hippopotamus, or hippo, and both are found in tropical Africa. The common hippo is one of the largest land mammals after the elephant, but the pygmy hippo is much smaller.

Hippos feed on plants, and live in or near rivers and lakes. They are well suited to life in water. They have nostrils on top of their snouts that can close, which stops water getting up their noses, and spread-out toes to help them swim.

Common hippos have bulky bodies and huge heads and mouths, with large teeth like small tusks. They usually live in groups of 15 or so, but sometimes many more herd together. Common hippos spend most of the day resting in water.

Often they only leave their ears, eyes, and nostrils above the surface. They are good swimmers, and can walk along the bottom of rivers and lakes. At night they come on to the land, where they feed on grass and other plants.

Hippos have grey, almost hairless skins. The pinkish appearance of many hippos comes from a special oil that they produce to protect their skins. In the past, this led people to think that hippos 'sweat blood'.

▼ Common hippos can weigh up to 4.5 tonnes, and be over 3 m long. Pygmy hippos are less than 2 m long, and spend more time on dry land than common hippos. They live in rainforests, and are now very rare.

find out more
Mammals

• Hippos are related to rhinos and horses. The word 'hippopotamus' means 'river horse'.

Margin notes provide nuggets of extra information and amazing facts.

Colourful illustrations and photographs bring the topic to life.

Boxes highlight records, statistics and amazing facts, or cover a particular aspect of the topic.

Horse records
Largest
One Percheron stood 21 hands (over 2 m), and measured over 5 m long
Smallest
The Falabella: adults about 70 cm (7 hands)
Fastest
69.2 km/h averaged over 400 m

• There are seven different kinds of horse, of which two (donkeys and horses) have been domesticated.

• A pony is a small horse, standing less than 142 cm (14 hands) at the withers (the high-point between its shoulder-blades).

find out more
Grasslands
Mammals
Zebras

The **find out more panel** points you to other articles related to the topic.

Horses

For thousands of years, and in almost all parts of the world, people have used horses for riding, to pull heavy loads and for sport and leisure. Herds of wild horses still survive in some parts of the world – though all but the zebras of Africa are very rare.

Wild horses live in herds in open grassland, in Africa, the Middle East and Asia. They have sharp senses of sight, hearing and smell, and can detect hunting animals from a distance. They are strong runners, and use their speed over long distances to escape from danger. Herds of feral horses (domestic horses gone wild) survive in many places, especially in the Americas and Australia.

Domestication of horses

Horses were first domesticated by prehistoric people in central Asia over 6000 years ago. At first, they were used to pull lightweight chariots, but when larger breeds were developed, people began to ride. At first they rode without saddles or stirrups. The Roman cavalry were the first to use saddles regularly, and stirrups were only

introduced into Europe from Asia in the 9th century AD. With the arrival of the horse collar at about the same time, people began to use horses to pull heavy weights.

During the Middle Ages, the armour worn by knights was so heavy that only very big horses could carry their weight. More lightweight, speedier horses were bred in the countries surrounding the Mediterranean Sea. These were the ancestors of the fast, slender-limbed horses such as the Arab breeds.

▲ The small, stocky Przewalski's horse comes from the steppes (grasslands) of Mongolia and western China. Herds are found in zoos and reserves, but it is probably extinct in the wild.

Captions not only describe the photographs and illustrations but give additional information on the topic.

The footer provides a short cut to topics that do not have their own articles.

5

Algae

Algae are plants without leaves, stems, roots or flowers. They are found in wet places everywhere, from oceans and rivers to the damp side of a tree. Algae can use sunlight to make food (photosynthesize) like higher plants, but strictly they do not belong to the plant kingdom.

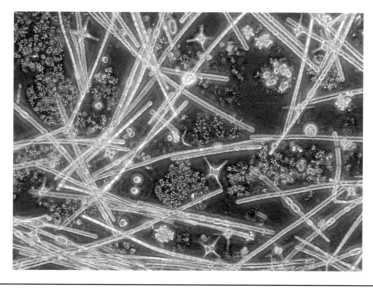

Many kinds of algae are microscopic. When millions grow together in one place, they look like green slime. Such small algae are found in lakes and rivers, and most importantly in the sea.

Floating food

Most of the plant life of the oceans is tiny, single-celled algae. They are known as *phytoplankton* ('plant plankton'). Plankton is the name given to the billions of tiny, floating organisms that live near the surface of the oceans. They form the basic food of many of the animals that live in the ocean. One type of algae among the phytoplankton are the *diatoms*. They do not look like land plants, but have a silica shell that protects the living plant inside. They cannot swim, but many of them have strange, spiky shapes that prevent them from sinking.

When algae grow in huge numbers, they are called *algal blooms*. These cause great harm to sea life, for they are usually poisonous to animals.

Algal sandwiches

Lichens are inconspicuous plants that often form black, orange or greenish-grey patches on rocks, walls and roofs. A lichen is like a sandwich of a fungus and an alga. The alga makes the lichen's food from sunlight and air. On the outside, strands of fungus protect the algal filling.

▶ Some types of phytoplankton, magnified many times. The diatoms can be recognized by their spiky shapes.

find out more
Fungi
Oceans and seas
Plants
Ponds
Seaweeds

Amoebas

Amoebas belong to a group of tiny animal-like creatures called *protozoans*. Protozoans have only one cell, and can only be seen through a microscope. Although tiny, this single cell is quite complicated, and can do many of the things that larger animals and plants can.

Amoebas are continually changing shape. This is the way they move, and the way they eat. To move forward, an amoeba makes a bulge in its body called a 'false foot'. This attaches to a surface, and pulls the rest of the amoeba forward. The amoeba then makes a new false foot, and repeats the movement. To eat, the amoeba moves towards a tiny particle of food, then changes its shape so that it is surrounding the particle. The food is then taken inside the amoeba and digested. An amoeba reproduces simply by splitting into two separate individuals.

Many kinds of amoeba live in streams or ponds, and are harmless. However, other kinds of amoeba live as parasites in the guts of animals, including humans, and cause diseases.

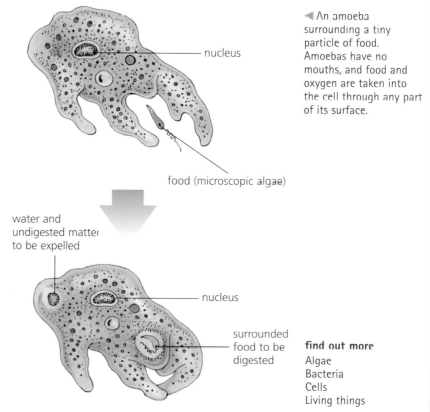

nucleus

food (microscopic algae)

water and undigested matter to be expelled

nucleus

surrounded food to be digested

◀ An amoeba surrounding a tiny particle of food. Amoebas have no mouths, and food and oxygen are taken into the cell through any part of its surface.

find out more
Algae
Bacteria
Cells
Living things

Alligators *see* Crocodiles and alligators

Amphibians

Frogs, newts and blindworms are amphibians. Amphibians are animals that start their lives in water and change as they grow up, so they are able to live on land as well. All amphibians are vertebrates (animals with backbones) and most have moist, soft skins.

- There are 3500 different kinds of frog and toad, 300 different kinds of newt and salamander and 167 different kinds of blindworm.

- Amphibians were the first vertebrates to live on land. About 400 million years ago there were fishes that had lungs and strong fins. When the pools where they lived dried up, they probably pulled themselves along on their fins. Gradually they became less dependent on water.

- The word 'amphibian' comes from two Greek words and means 'both ways of life'.

find out more
Frogs and toads
Newts and salamanders
Prehistoric life

Most amphibians are small animals and, to protect themselves from predators (animals that eat them), many hide during the daytime and are active at night. Many kinds of amphibian have poison glands in their skin which make them taste so nasty that hunting animals avoid them.

Amphibians have very small lungs and breathe partly through their skin. But skin can only breathe if it is kept damp, and so amphibians tend to live near water.

Amphibians are 'cold-blooded' animals. This means that their body temperature and activity depend on the warmth of the air or water around them. When the weather is warm they can be active. When it is cold, they become very sluggish. In cooler parts of the world, such as northern Europe, amphibians hibernate through the winter months.

Metamorphosis

The life of an amphibian begins in water. As it grows up, it changes to adapt to life on land. The process of change is called *metamorphosis*. Most adult amphibians lay large numbers of eggs, or spawn, in ponds or streams. The eggs hatch quickly, but the baby that emerges looks very different from its parents. As it grows, its body changes. It develops lungs and legs, so it can live on dry land. Its diet changes too, and it begins to feed on tiny creatures. Adult amphibians eat flesh, feeding on many sorts of small animals.

Kinds of amphibian

There are three main kinds of amphibian. Blindworms are strange creatures found only in the tropics. They have no legs and a grooved skin, like earthworms, and they burrow

▲ Blindworms are unusual amphibians because only one species lives in water.

through the ground feeding on worms and grubs. Some young blindworms hatch from eggs and some are born live.

Newts and salamanders look like soft-skinned lizards. They are often brightly coloured. Some kinds, such as axolotls, never leave the water. Other kinds crawl up onto land and live in damp woodland areas. One sort hides in logs that people use to make fires, and so it is called the fire salamander.

Frogs and toads have no tails, but they usually have long, powerful hind legs, which they use for jumping and swimming. There are far more kinds of frog and toad than of any other amphibian. They are an important part of the life of tropical forests.

The life cycle of a frog

Frogs' eggs have no shells. The outside of the egg swells to form a jelly-like protective covering.

A baby frog (*tadpole*) looks rather like a little fish, with a big head and a wriggly tail.

At first the tadpole breathes with gills and feeds on tiny plants that live in the water.

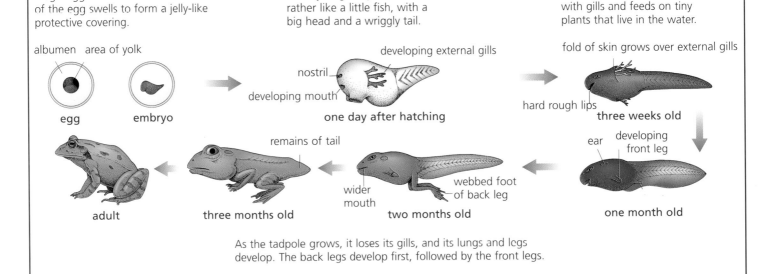

albumen   area of yolk

egg   embryo

nostril
developing mouth
developing external gills
one day after hatching

fold of skin grows over external gills
hard rough lips   **three weeks old**

ear   developing front leg
webbed foot of back leg
one month old

remains of tail
wider mouth   **two months old**

three months old

adult

As the tadpole grows, it loses its gills, and its lungs and legs develop. The back legs develop first, followed by the front legs.

Animal behaviour

An animal's behaviour includes everything that it does. It may search for food and eat, it may try to escape from its enemies, and it may find a mate and produce and rear young. Even sleep is part of its behaviour.

Most animal behaviour can be broken down into a series of separate actions caused by changes in its circumstances. These changes, which are called triggers, may be internal. For instance, when an animal has digested a meal it feels hungry. This is the trigger for it to search for more food. Its senses pick up external triggers and tell it what is suitable food. For instance, a cheetah in Africa may start to hunt because it is hungry. It will see but ignore large antelopes, but the sight of a Thomson's gazelle will be the trigger to start the chase.

Instinct and learning

The behaviour of many animals looks very clever, but often the cleverness is something that is automatic and cannot be changed. For instance, wasps could no more alter the way they build their beautiful, complex paper nests than they could change their yellow and black colours. Wasps do not have to learn how to make their nests. There are many examples of such unlearned activities, which are often called *innate* (or *instinctive*) *behaviour*.

▼ Like all spiders, this orb-weaver spider from Trinidad in the Caribbean is born knowing how to build the complex web that it uses to catch food.

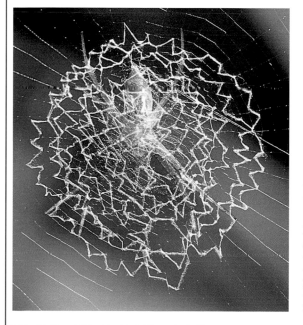

On the other hand, many creatures learn from experience and can change their behaviour. All vertebrates (backboned animals) and a few invertebrates, such as octopuses, behave in ways that are partly learned. An animal's actions are often a complex mix of instinctive and learned behaviour. For instance, all young animals are born with a knowledge of how to eat, but they often do not know what to eat, and have to find out by trial and error. A chick may peck at all sorts of things, such as small stones, but once it has found the right food it does not try to eat inedible things anymore.

Animal language

Many kinds of animal live in groups and need to be able to communicate with each other. Each kind of animal has its own language, although warning noises may be understood by other species. Cats and dogs that share a home do not share a language.

Animals communicate in many ways. Many give and receive information by scent, for instance by using it to mark the boundaries of their territory. Animals also use body language, as when they display specially coloured areas, such as crests or tails, to put across a simple message. When animals are close together, they can communicate using facial expressions. These are important in many mammals. Watch dogs playing and see how they open and narrow their eyes and alter the position of their ears and lips to express meaning. Some animals use sound language. Apes and some kinds of bird are known to have at least 25 sounds with exact meanings.

▲ This chimpanzee is using a stick to dig termites out of a termite mound in Zambia. It has probably learned to do this by copying other chimpanzees, and then trying the behaviour out for itself.

• Many of the things that humans do are innate. We respond to triggers just as other creatures do, but more of our behaviour is learned and our language is hugely complex.

• In the 1950s the Austrian naturalist Konrad Lorenz showed that goslings will follow anything they see during their first 16 hours after hatching. Following is instinctive but what they follow is learned. When introduced to his boots during this 'critical period', young geese followed Lorenz's booted feet everywhere.

find out more
Animals
Hibernation
Migration

Animals

Animals live almost everywhere in the world, from polar regions to hot, dry deserts and steamy forests. The biggest weigh over 100 tonnes, the smallest can only be seen through a microscope.

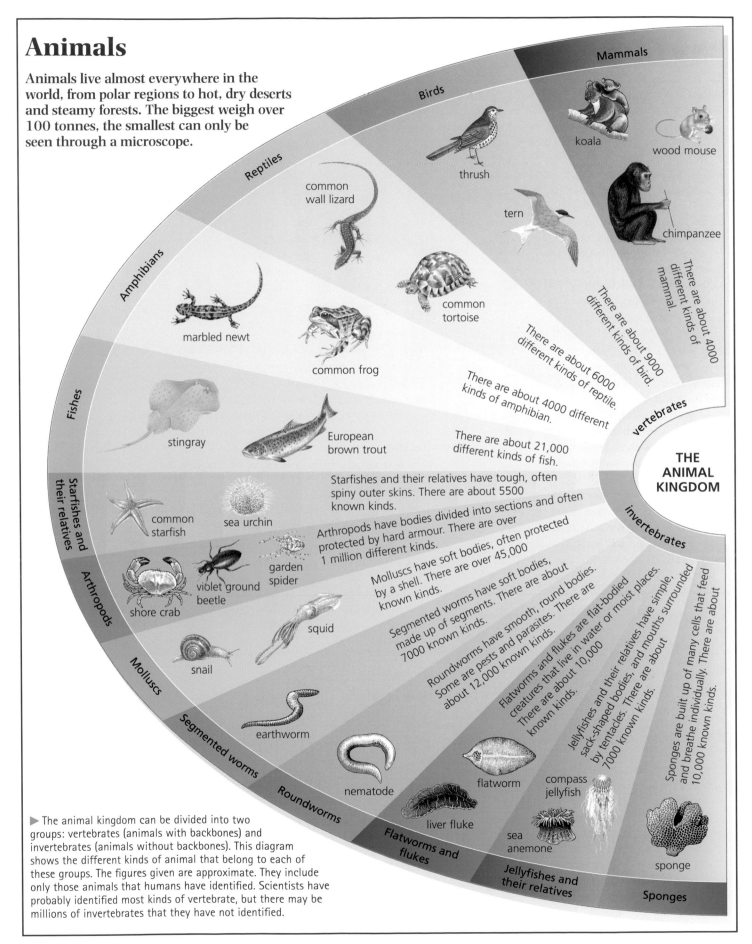

Mammals

koala

wood mouse

chimpanzee

There are about 4000 different kinds of mammal.

Birds

thrush

tern

There are about 9000 different kinds of bird.

Reptiles

common wall lizard

common tortoise

There are about 6000 different kinds of reptile.

Amphibians

marbled newt

common frog

There are about 4000 different kinds of amphibian.

Fishes

stingray

European brown trout

There are about 21,000 different kinds of fish.

vertebrates

THE ANIMAL KINGDOM

invertebrates

Starfishes and their relatives

common starfish

sea urchin

Starfishes and their relatives have tough, often spiny outer skins. There are about 5500 known kinds.

Arthropods

shore crab

violet ground beetle

garden spider

Arthropods have bodies divided into sections and often protected by hard armour. There are over 1 million different kinds.

Molluscs

snail

squid

Molluscs have soft bodies, often protected by a shell. There are over 45,000 known kinds.

Segmented worms

earthworm

Segmented worms have soft bodies, made up of segments. There are about 7000 known kinds.

Roundworms

nematode

Roundworms have smooth, round bodies. Some are pests and parasites. There are about 12,000 known kinds.

Flatworms and flukes

flatworm

liver fluke

Flatworms and flukes are flat-bodied creatures that live in water or moist places. There are about 10,000 known kinds.

Jellyfishes and their relatives

compass jellyfish

sea anemone

Jellyfishes and their relatives have simple, sack-shaped bodies, and mouths surrounded by tentacles. There are about 7000 known kinds.

Sponges

sponge

Sponges are built up of many cells that feed and breathe individually. There are about 10,000 known kinds.

▶ The animal kingdom can be divided into two groups: vertebrates (animals with backbones) and invertebrates (animals without backbones). This diagram shows the different kinds of animal that belong to each of these groups. The figures given are approximate. They include only those animals that humans have identified. Scientists have probably identified most kinds of vertebrate, but there may be millions of invertebrates that they have not identified.

Animals are one of the five main groups (called kingdoms) into which all living things are divided. It is not always easy to tell animals from plants, but generally animals are able to move about from place to place and must eat food to survive.

All animals eat things that have once been alive. Most, like horses and snails, are *herbivores*, and eat plants. Some, like tigers and sharks, are *carnivores*, and kill and eat other animals. Some, like human beings and worms, are *omnivores*, and eat almost anything. Others, like vultures and dung beetles, are *scavengers*, and feed on dead and rotting plants and animals.

Vertebrates and invertebrates

The word 'vertebrate' comes from the Latin word for a backbone, so vertebrates are animals with backbones. The bodies of vertebrates are supported by the backbone and the other bones in the skeleton. There are five major groups of vertebrates: fishes, amphibians, reptiles, birds and mammals.

Almost all the bigger, more complicated animals are vertebrates, but there are many more different kinds of invertebrates – animals without backbones. These include insects, squids, snails, jellyfishes and worms. Invertebrates do not have bones but may be supported by some kind of hard shell outside or inside their body. There are about 50,000 different kinds of vertebrate that we know of. This is probably less than 1 per cent of all the animals that exist.

Activity

Some animals – birds and mammals – are 'warm-blooded'. This means they are able to keep their body at a constant temperature, however hot or cold their environment. They are always able to be active, but they pay a high price as they have to find a great deal of food to fuel their activity. However, most animals, including reptiles, amphibians, fishes and all invertebrates, are 'cold-blooded'. This means that their body temperature rises and falls with the temperature of their surroundings. In sunshine or hot weather, they can be quite warm and active, but at night or in winter, they are often cold and inactive. The advantage to being so inactive is that they do not need so much food.

Reproduction

Most animals are either male or female, and after mating, the young are born sharing the characteristics of both parents. A small number of animals, such as greenfly and water fleas, are almost all females and, without mating, are able to produce many female young. A few kinds of animal, such as garden snails, are both male and female at the same time, and are able to produce both eggs and sperm. A few others, such as oysters and some fishes, change sex at a certain time in their lives or as a result of changes in their environment. Others still, like some sea anemones, are able to divide their own body, splitting off pieces that grow to make the next generation.

Some animals produce huge numbers of young, but most of these die early, eaten by other creatures. Animals that have long lives and that care for their young tend to have smaller families. Usually the female animal is most important in caring for her family, but in some cases the male helps equally and with some animals he does all the work of looking after the offspring.

Animal records

The *largest* animal is the blue whale, which measures about 30 m in length and over 100 tonnes in weight.

The *tallest* animal is the giraffe, which grows up to 5.3 m tall.

The *largest land* animal is the African elephant. It stands up to 3.2 m at the shoulder and weighs up to 8 tonnes.

The *fastest land* animal is the cheetah, which can run, for a short distance, at up to 100 km/h.

Anteaters *see* Armadillos, anteaters and sloths

Antelopes

Most antelopes live in large herds on the grasslands of Africa, where they feed on a wide variety of plants. Antelopes are swift runners, and have long legs, which allows them to avoid predators such as lions. Some antelopes, such as the royal antelope, are tiny, while many, such as the wildebeest (or gnu), are big, powerful animals.

Often a herd of antelopes consists of females and calves only, although males will join a herd

during the mating season. Males fight other males for females and for territory. Their horns are usually spiral or notched so that the rivals' horns lock together when fighting. There follows a trial of strength as they push each other backwards and forwards. Although the fights look fierce, the loser can easily break free and injuries are rare.

A baby antelope can stand within minutes of its birth and is soon an active member of the herd. It becomes independent quickly, as its mother will produce a new calf in the next year.

Antelopes are the prey of many flesh-eating animals, such as lions and other big cats, but humans are their most dangerous enemy. Many species have now become very rare because of over-hunting or the destruction of their habitat. Attempts have been made to domesticate some kinds of antelope, and it is possible that they may become important farm animals in the future.

◄ This group of animals includes three different kinds of antelope: eland (the large animal to the right), waterbuck (medium-sized and dark brown) and impala (small, reddish-brown). Antelopes graze and browse on a wide variety of plants, which means that many species can live side by side without competing for food.

Ants and termites

Ants and termites are insects which always live in groups, or colonies. Most ant and termite colonies build huge nests. There are many thousands of individuals in a colony, but they all have the same mother: the queen.

Termites eat only plant food, and because of their numbers, they can become pests. Ants feed on many kinds of food, including plants and other animals. Ants can be found worldwide, while termites live only in warm regions, especially Africa, Australia and America.

Most ants and termites do not have wings, but at certain times of the year huge numbers take to the air. After a brief mating

flight, the females make new nests and lay eggs. Male ants die after mating, but male termites survive to share the royal chamber with the queen.

The eggs laid by the queen ant hatch into female ants called workers, whose job it is to collect food for themselves, their sisters and the queen. These are the ants that you are most likely to see. When they are out getting food, they make scent trails so that they can find their way back to the nest. Some of the workers become soldier ants, growing large and helping to protect the nest by biting or squirting acid at enemies.

When young termites hatch out, they look like miniature adults, but they do much of the work of the nest.

▼ Some kinds of termite build enormous nests, which may rise many metres above the ground. They contain fungus gardens where they grow food, cells for eggs and young, and the royal chamber in which the queen and her mate live.

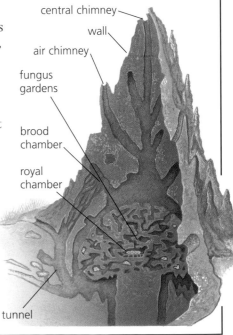

central chimney
wall
air chimney
fungus gardens
brood chamber
royal chamber
tunnel

Apes

Apes are our nearest relatives in the animal kingdom. Like humans, they are intelligent, long-lived mammals, and they generally live and travel in family parties. There are two main groups of apes. Gorillas, chimpanzees and orang-utans are the great apes, and gibbons are the lesser apes.

The great apes have bodies much like ours. Like us they have no visible tail, and 32 teeth. Vision is their most important sense, and unlike most mammals, they can see colours much as we can. Female apes, like humans, normally have one baby at a time. Young apes develop slowly. A mother does not have another baby until the last one is several years old. A baby chimpanzee continues to feed on its mother's milk until it is about 4 years old. Young chimpanzees learn many things from their mothers, such as how to use a twig as a tool to fish termites out of their nests.

Apes eat mainly leaves and fruit, though some also feed on flesh, such as insects. They wake early and roam through the forest in search of food. During the middle of the day they doze in nests of branches and leaves which they build in the trees. They dislike rain, and sometimes protect themselves by holding a leafy branch over their heads.

▼ Chimpanzees live in the forests of central and western Africa. They use gestures and facial expressions to give information to other members of the group, and have a language that includes at least 24 sounds.

• There are 13 kinds of ape: 9 lesser apes and 4 great apes.

• Great apes live about 50 years, while lesser apes live 30 to 40 years.

• Like humans, apes are primates. This means that their hands have a thumb which can be pressed against a finger to pick things up, although they cannot fold it across the palm of the hand. Unlike humans, apes also have long toes and can grasp things with their feet.

▲ Gibbons live in South-east Asia. They are the smallest of the apes. The largest gibbon, the siamang, grows up to 90 cm. Gibbons are not aggressive, but warn their neighbours away from their territories by loud calling.

Apes and humans

There are some obvious differences between apes and humans. The apes all have far more hair covering their bodies than we do. Their arms are much longer and stronger, while their legs tend to be shorter and weaker. Apes can climb very well and often swing about in the branches of trees. They usually walk on all fours. This is known as knuckle-walking, as they put their weight on their knuckles.

In spite of the fact that apes are our close relatives, we have not treated them well. We have killed them, imprisoned them in zoos and experimented on them in laboratories. Worse, much of the forest in which they live has been destroyed. The orang-utan, gorilla and gibbon are all in danger of dying out, and chimpanzees are much less common than they used to be.

▼ Gorillas are the largest of the great apes, growing up to 1.75 m tall. They live in the forests of central Africa. Gorillas are unaggressive creatures, and live in family groups led by a big male, the silverback.

▲ The word 'orang-utan' means 'man of the woods'. Orangs live in forests on the islands of Sumatra and Borneo in South-east Asia. Unlike the other great apes, they mostly live alone.

find out more
Animal behaviour
Human beings
Mammals
Monkeys
Primates

Armadillos, anteaters and sloths

• Armadillos, anteaters and sloths are all found in South and Central America; the nine-banded armadillo is also found in parts of North America.

• There are 20 different kinds of armadillo, four kinds of anteater and five kinds of sloth.

• Armadillo means 'little armoured one'.

find out more
Ants and termites
Mammals

Armadillos, anteaters and sloths all belong to a group of mammals known as edentates. This name means 'without teeth', but only the true anteaters are totally toothless.

Armadillos have plates of bone and horn on their backs, heads and tails. They are mainly active at night, when they search for insects, worms and carrion (dead animals). If attacked, some kinds curl up into a ball, while others defend themselves with claws on their front feet. But they are not aggressive, and prefer to escape danger by burrowing underground.

Anteaters eat thousands of ants and other insects a day. Giant anteaters live on the ground; other kinds of anteater live in trees, using their tails to cling onto the branches.

Sloths live in tropical trees, and use their huge, curved claws to hang upside-down from the branches. They rarely come down to the ground, where they are nearly helpless. Sloths have long coarse hair on which tiny plants (algae) live.

▶ Sloths are divided into two groups: one group has three toes, and the other has two toes. This is a three-toed sloth.

◀ As the nine-banded armadillo walks, the only part of its front feet to touch the ground are its large claws.

▲ The giant anteater's tongue is covered with tiny backward-pointed spines. These become sticky with saliva, making it easy to mop up lots of insects.

Bacteria

Bacteria are very simple living things, consisting of a single cell. A row of a hundred would just reach across this full stop. Bacteria are found everywhere: in the ocean depths, in soil, on our skins, even floating freely in the air. They are some of the oldest living things, and fossils have been found of bacteria that lived 3000 million years ago.

Some bacteria can make their own food by using the Sun's energy (photosynthesis), but most live on decaying plants and animals, or as parasites in other living things.

By breaking down dead plants and animals, bacteria return vital nutrients to the soil, and these are used by plants for growth. But they also cause some of the deadliest diseases of humans, such as tuberculosis, pneumonia, cholera and typhoid.

Unlike single-celled animals and plants, bacteria have no nucleus or distinct parts. They do not have sexes either, but reproduce by dividing in two. They do this so quickly that, given enough food, one bacterium could produce 4000 million million million others in just one day.

▶ **FLASHBACK** ◀

People once believed that old food went bad because it changed into moulds and microbes (bacteria). In the 19th century the French scientist Louis Pasteur showed that it was microbes in the air that caused the decay rather than a change in the food. He placed freshly cooked food in clean glass containers, sealed one and left the other open to the air. Only food in the open container went bad.

find out more
Cells
Forests
Living things
Pests and parasites

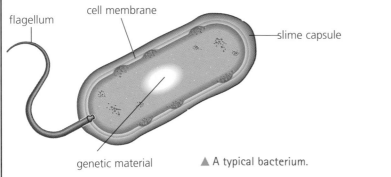

flagellum
cell membrane
slime capsule
genetic material
▲ A typical bacterium.

Baboons *see* Monkeys • **Badgers** *see* Weasels, etc. • **Bamboos** *see* Grasses

Bats

- There are over 950 kinds of bat, about a quarter of the total number of mammal species.

- There are three kinds of South American vampire bat, which feed only on the fresh blood of mammals and large birds. They do not suck blood, but lap it up. More dangerous to the victim than blood loss is the fact that vampire bats can carry diseases, including rabies.

find out more
Cave animals
Ears
Hibernation
Mammals
Migration

Bats are the only mammals that have wings and are capable of true flight. During the daytime they are rarely active, resting in dark, hidden places such as caves, lofts and holes in trees. When they emerge, bats are able to find their way in total darkness and silence.

Bats are nocturnal (active at night), and find their way in the dark using *echolocation*. As it flies, a bat makes short, very high-pitched squeaks. When something is nearby, echoes from the squeaks bounce back to the bat's ears, giving it details about the object, such as its distance, size and shape.

Flight requires a lot of energy, so bats eat a large number of night-flying insects. Some tropical bats feed on larger prey, including other kinds of bat and small creatures such as mice and frogs. Others feed on nectar and fruit.

To conserve energy when they are not active, bats let their body temperature drop. In seasons when food is scarce, bats living in cool regions hibernate (go into a deep sleep), but a few kinds migrate to warmer regions where food is plentiful.

Bats are hunted by some animals, but humans are their main enemy. Some kinds are becoming rare due to pollution and destruction of their homes.

◀ The front part of a bat's wing is supported by the bones of its arm. The wing is also braced by very long finger bones. This fishing bat catches fish with its very large hind feet, which have specially long, strong claws.

Bears

- There are eight different kinds of bear. The largest is the Alaskan Kodiak bear (a kind of brown bear), which can grow to 3 m in length and weigh up to 780 kg.

find out more
Conservation
Hibernation
Mammals

▶ A female grizzly (or brown) bear with her three cubs. There used to be many brown bears in Europe and North America, but much of the land they lived on is now used for farming. Today they are mostly found in Alaska, Canada and the remoter parts of Asia.

Bears are huge creatures: the biggest bear weighs more than three times as much as a large lion. They are found in a wide range of habitats, from the frozen Arctic to tropical forests. The polar bear feeds mainly on fish and seals, but other bears generally have a varied diet, including fruit, nuts, honey and small animals.

Like human beings, bears can stand upright, but normally they walk on all fours. They have poor eyesight and hearing, but their sense of smell is excellent. They generally live alone. In the autumn, when food is abundant, bears eat a lot before sleeping in a den for much of the winter. However, they do not go into a deep hibernation (winter sleep) and are usually able to wake up quickly.

A female bear produces her cubs in the wintertime, and they stay with her in her den until springtime. The cubs are very small, rarely larger than guinea pigs. They remain with their mother until they are more than a year old. When they leave her, the cubs usually stay together – often for as long as three years.

The giant panda is a relative of the true bears (although the other kind of panda – the red panda – is more closely related to racoons). It is one of the most striking-looking animals in the world, and also one of the rarest: there are only about 1000 left in the wild. Pandas live in the high mountains of only three isolated parts of China. Bamboo is their main diet, although they do eat other plants, and occasionally hunt small birds and mammals.

Beavers *see* Mice, squirrels and other rodents

Bees and wasps

Bees and wasps are amongst the most familiar insects. Many live alone, but some, such as honeybees and bumble-bees, live in family groups in a hive or nest.

Each hive or nest is the home of a large number of bees and their mother, the queen bee. Worker bees are females that look after the grubs that hatch

from the queen bee's eggs. A small number of the bees in a hive are males, or drones. Their job is to mate with young queen bees, after which they die.

Honeybees make honey from a sugary liquid called nectar, which they suck from flowers. The bees use it to feed their young and as food during the winter, when there are no flowers to provide nectar.

As a bee searches flowers for nectar, she also collects pollen from the flowers into 'pollen baskets' on her hind legs. She

takes the pollen back to the hive, where it is used as the main food for the grubs. Some of the pollen also gets stuck on her coat, and it gets brushed onto the next flower she visits, and pollinates (fertilizes) it.

Wasps are closely related to bees. Most people think of wasps as insects with yellow and black stripes that protect themselves with a sting. However, these are social wasps, which are only one of the many different kinds of wasp.

- There are about 25,000 different kinds of bee and over 50,000 kinds of wasp.

- A bee's sting is formed from the egg-laying tube of the female. It is like the needle of a hypodermic syringe and pumps poison from a gland at its base into the enemy. If the sting is used against a human or other mammal, it gets stuck in the skin and the bee dies.

▶ Worker bees at work.

find out more
Animal behaviour
Insects

Workers follow the queen when she lays eggs. She produces a chemical called 'queen substance', which attracts the other bees.

Workers construct the comb from wax made in their bodies. It may be used as a store for honey, or as a nursery for the grubs.

A worker visits a flower to collect nectar and pollen. Bees are the most important pollinators of flowers.

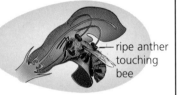

ripe anther touching bee

Beetles

There are more different kinds of beetle than any other sort of animal. They are easy to recognize because they look as if they are wearing heavy armour. In fact, their front pair of wings are thickened and hinged down to cover the body. Their hind wings are large and papery.

Although there are thousands of different beetles, we do not often see very many. This is because most of them hide away during the daytime, or else keep busy under fallen leaves or among the roots of grasses.

Beetles start life as an egg, and then hatch into a larva (grub). Unlike its parents, it has

a soft body, but it has hard biting mouthparts to help it feed and grow. When it has finished growing, the grub finds a safe place to pupate: it goes through a resting stage (pupa), during which the fat body of the larva changes to form the adult beetle. This dramatic change is called metamorphosis.

A few beetles are pests because they feed on crops or food stores. However, many more are useful. Some are recyclers, helping to turn decaying matter into nutrients

for the soil, which in turn help plants to grow. Others, such as ladybirds, feed off insect pests such as aphids which eat plants, and so are useful to farmers and gardeners.

- Ladybirds are small beetles. They are usually brightly coloured, with dark spots on an orange, yellow or red background. Each kind of ladybird has a different pattern and number of spots.

- The Goliath beetle is the biggest beetle, as well as the biggest insect. It is about 15 cm long and weighs about 50 g.

find out more
Insects

3

2

4

▲ Scientists have discovered over 400,000 different kinds of beetle, and new ones are being found all the time. The ones shown here are a violet ground beetle (1), which cannot fly; a tiger beetle (2), which feeds mainly on other insects; a great diving beetle (3), which lives in water; and an oil beetle (4), which lives in the desert.

Biology

Biology is the study of living things and the ways in which they interact with the world around them. Biologists study everything about the living world, from the workings of minute bacteria, to the evolution (change) that occurs in animals and plants over millions of years.

Biology is a very broad subject, and it can be studied in many different ways. Some biologists look at large groups of living things. *Population biologists* study whole populations of animals or plants, while *taxonomists* try to classify the different kinds of living things into groups, depending on how closely they are related. *Ecologists* look at how different animals and plants in the same area interact with each other and with their environment.

Other areas of biology look at animals and plants on an individual level. *Anatomists*, for example, study the structure of individual plants or animals, while *physiologists* look at what the different parts do and how they work. Some biologists become experts on particular types of

◀ Some students of biology go on to become veterinarians (vets). In the same way that doctors know all about human biology, vets are expert in animal biology. This vet is applying medicine to the eye of a white rhino, an endangered animal, in Kenya.

living thing. *Zoologists* study animals; *botanists* study plants. *Ornithologists* study only birds, while *entomologists* specialize in insects.

The area of biology that has grown most in the last 100 years is the study of living things at the microscopic level. All living things are made up of tiny cells, and *cell biologists* look at the different kinds of cell and how they work. *Molecular biologists* work on an even smaller scale, studying the chemicals (molecules) that make up cells. These complex molecules, and the chemical reactions that occur between them, are basic to all life on Earth.

find out more
Animals
Cells
Ecology
Evolution
Genetics
Living things
Plants

Biotechnology

• Genetic engineering can also be used to make 'improved' plants. Food plants such as wheat and rice can be given genetic material from other plants to improve their resistance to disease or to help them grow in difficult conditions. Some people are worried that we do not know enough about the effects of altering plants in this way. They think that other plants, and whole ecosystems, may be damaged by such human interference.

find out more
Bacteria
Fungi
Genetics

Biotechnology is the use of living things to make large amounts of useful products such as drugs, vaccines and medicines. It often involves the use of microbes – tiny single-celled creatures such as bacteria and yeasts.

Microbes have been used in the production of some foods for centuries. Bakers and brewers use microbes called yeasts to make bread and beer. Cheese-making involves the use of bacteria and fungi.

Most modern biotechnology relies on a process known as *genetic engineering*, in which scientists alter the cells of simple animals and plants to give them useful properties. An example of this is the production of human insulin.

Insulin is a chemical that controls the levels of sugar in the blood. People with the disease diabetes need insulin as part of their treatment. To make insulin, scientists take a few human cells and extract from them the genetic material (DNA) that has instructions for

making insulin. They copy this material many times, and insert it into microbes, which then produce insulin.

Biotechnology has many other uses in medicine. Microbes can be genetically engineered to produce new drugs and vaccines. They are also used to make human antibodies, substances normally produced in the body which fight disease.

Other products that can be made include enzymes. These are biological substances that speed up chemical reactions in the body. They have many uses in medicine and industry.

◀ Once a microbe has been modified to produce a substance, large numbers have to be grown. Colonies are first grown on flat dishes containing food, as with these fungi. Later, huge numbers are grown in large containers.

Birds

Birds live in all parts of the world: rainforests, deserts, oceans and even the icy wastes of Antarctica. Up in the sky, swans may migrate at a height of 8000 metres. Under the water, penguins may swim to a depth of over 250 metres. There is great variety amongst birds. Some birds eat only plants and some eat only fresh meat. The ostrich grows to be taller than humans, while the bee hummingbird is small enough to fit in the palm of your hand.

Birds are warm-blooded animals. They are the only creatures that have feathers, and all have wings and lay eggs on land. Their bodies are specially designed to enable them to fly efficiently, although a few, such as kiwis and penguins, have lost the power of flight.

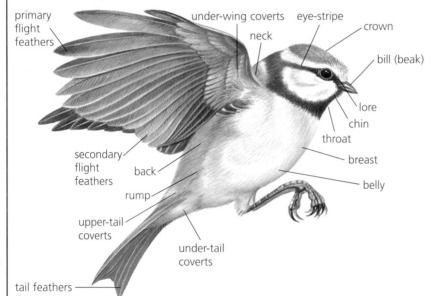

▲ Parts of a bird's body. The upper parts start at the crown and end at the upper-tail coverts. (A covert is a feather covering the base of a bird's tail and flight feathers.) The under parts begin with the chin and finish with the under-tail coverts.

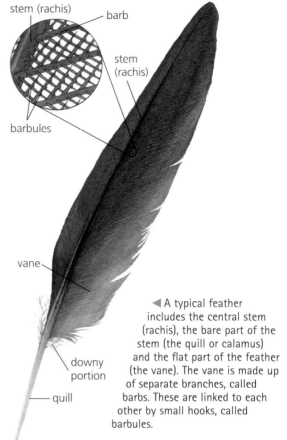

◀ A typical feather includes the central stem (rachis), the bare part of the stem (the quill or calamus) and the flat part of the feather (the vane). The vane is made up of separate branches, called barbs. These are linked to each other by small hooks, called barbules.

Feathers cover almost all parts of a bird's body. They are made by the bird's skin, in much the same way that our skin makes hair or fingernails. Usually, only a bird's beak, eyes, legs and feet remain bare.

Feathers have three other important functions. First, they are very good at keeping the bird dry and warm. In cold weather birds fluff up their feathers so that layers of insulating air are trapped between them. Second, the wing feathers enable birds to fly. The flattened curved shapes of the secondary wing feathers help provide the lift that keeps a flying bird in the air. The third important function of feathers is to give birds a streamlined shape, which also aids flight. Feathers can also be used for camouflage or to make a bird obvious. Females are often camouflaged, because they need to stay hidden while sitting on their eggs. Males are often showy and brightly coloured in order to attract a female.

Flight

Birds are specially adapted for flight. Their bones are very light and some are hollow. Their feathers are also light but they serve as a strong covering for the body. In the breeding season females lay eggs at regular intervals and the young develop in a nest so that the mother is not weighed down by carrying them.

Not all birds fly in the same way. Starlings fly in a straight line, moving their wings all the time. Woodpeckers close their wings between flaps and 'bound' along. Larger birds such as gulls often glide, and vultures hardly move their wings as they soar for hours on end.

Some birds, such as kestrels, can hover in the air as they look for prey on the ground, while peregrines, which hunt other birds, can reach speeds of 180 kilometres an hour as they dive to catch their prey. Hummingbirds can also hover in flight while they gather nectar from flowers, their wings beating at an incredible rate of up to 100 times a second.

Food

▼ Birds' bills are often shaped to help them feed. The curlew (1) uses its long bill to probe mud-flats. Pelicans (2) use their elastic throat pouch as a type of fishing net. The flamingo (3) has a sieve-like bill for filtering food from the water. Eagles (4) tear meat with their hooked bills. The hawfinch (5) has a short, strong bill with which it can crack hard food such as cherry stones.

To avoid competing for the same food, different kinds of bird have adapted to different sorts of food. The shapes of their beaks and feet often give us a clue as to what they eat.

Many wading birds have long, strong bills for probing the mud in search of worms or shellfish. Birds of prey have hooked bills for tearing meat. Ducks have flattened bills for filtering water or mud. Sparrows and finches have short, strong bills for cracking seeds.

Some fish-eaters, such as herons, have long legs for wading in water. Others that feed on fish, such as cormorants, have shorter legs and webbed feet to help them swim, dive and chase fish underwater.

▼ Japanese cranes performing a complicated courtship dance. The male and female strut around each other with quick, stiff-legged steps and their wings half-spread. The speed of the dance quickens and the birds leap over 4 m into the air and drift down as though in slow motion.

▲ The Arctic tern has the longest migration route of any animal. Including the return trip, it can cover a distance of over 35,000 km in one year.

——— main migration route
----- alternative migration routes

Migration

Many animals, including birds, migrate: they move from one place to another and return again in a different season. While some birds do not travel far from where they hatch, others, especially those breeding in the northern hemisphere, fly hundreds or thousands of kilometres to spend the winter in warmer places where there is plenty of food.

Sometimes only part of a population will migrate. Some mountain birds travel only a few kilometres to lower, warmer levels, while others will move all the way to the coast.

The migrations of small birds can be very long. Wheatears are only a little larger than sparrows, yet each year some fly non-stop from Greenland to North Africa. The tiny ruby-throated hummingbird migrates from North to South America, a journey that includes an 800-kilometre crossing of the Gulf of Mexico.

Sea birds make even longer journeys. Some, such as puffins, go out to sea and spend the winter out of sight of land. Arctic terns fly from their northern breeding grounds to winter in the Southern Ocean, sometimes reaching Antarctica.

Birdsong

Some birds make noises with their bills. A few use their feathers. But most use songs and calls to communicate with other birds.

Calls are usually short and loud, and they may signal danger. They may also be used by courting birds or by young birds begging for food. Some calls help to keep flocks together in thick woodland, or when migrating at night.

It is usually only male birds that sing, to tell other males that they have chosen a place to nest and are prepared to defend the surrounding area (their territory), or to attract a female. ▶

Danger from humans

Young birds must learn to find food and shelter, to survive extreme weather and to avoid enemies. Even so, most will die in their first year. Normally, enough survive to produce the next generation and a balance is created. Unfortunately, humans have the power to upset that balance.

The dodo was a victim of human interference. It was a huge, flightless relative of the pigeon, which lived on the island of Mauritius in the Indian Ocean. Visiting sailors killed many dodos for food, and the cats, pigs, rats and monkeys that the sailors brought with them destroyed the dodos' eggs. By 1681 all the dodos were dead. Since then the great auk, the passenger pigeon and many other birds have also become extinct as a result of human activities.

Birds are also hunted for food and as a sport by some people. Game birds such as pheasants, partridges and grouse are commonly hunted in Europe and North America. However, the activity is managed so that their numbers are not threatened.

Although birds are increasingly being protected by laws throughout the world, damage still occurs. For instance, around the Mediterranean millions of migrating birds are shot or trapped by hunters each year.

The rapid growth of the human population and the huge demand for natural materials have resulted in the destruction of many of the places where birds live. Marshes are drained to make farmland, forests are cleared for timber or for space to build houses or to grow crops, deserts

increase in size because of over-grazing of pasture land by domesticated animals. Even global warming caused by human pollution will affect bird populations.

The origin of birds

Archaeopteryx (from the Greek meaning 'ancient wing') is the earliest known fossil of a bird. It is a 'missing link' between the dinosaurs and the birds as we know them today. The first fossil of *Archaeopteryx* was found in Germany in 1861. It was contained in rocks of the Upper Jurassic period, which are about 145 million years old. The fossil showed an almost complete skeleton, about the size of a pigeon. The creature was similar to small dinosaurs but also had feathers, which showed that it was partly like a bird.

▲ A flock of budgerigars at a waterhole in Australia's North West Territory. Budgerigars are native to Australia but now breed wild in parts of North America too. Like other kinds of parrot, budgerigars are popular as pets. This is partly because they are able to copy human speech.

find out more
Animal behaviour
Conservation
Ducks, geese and swans
Eggs
Hunting birds
Migration
Ostriches and their
  relatives
Sea birds
Songbirds
Wading birds

▶ Domestication involves controlling the living and breeding conditions of birds. Many different birds have been domesticated to provide meat and eggs.

turkey

Chinese goose

rooster

Indian runner

chicken

Blood

Blood is the body's transport system. It carries food and oxygen to all parts of the body and removes harmful wastes. It spreads heat evenly around the body, and plays an active part in fighting injury and infection.

In most animals, blood is pumped round the body by

◀ A coloured and greatly magnified photograph of blood. The red blood cells are coloured red. The white, rough-looking objects are white cells. The small blue objects are platelets, fragments of bone marrow which help the blood to clot.

some kind of heart (a muscular pump) and travels through tubes called blood vessels. In insects and many other invertebrates (animals without backbones) the blood vessels are *open-ended* – the blood flows through the spaces between the animal's cells. In some invertebrates and all vertebrates (animals with backbones) the blood flows through a system of closed vessels. Large vessels (*arteries*) branch into smaller and smaller vessels, carrying the blood away from the heart.

The blood of vertebrates is made up of a fluid called plasma, with blood cells floating in it. Red blood cells carry oxygen from the lungs (or the *gills* in fish) to all the other cells of the body. The blood also takes dissolved foods such as glucose (sugar) from the intestines (gut) to other parts of the body and carries away wastes such as carbon dioxide and urea.

Injury and infection

Blood can protect us from damage and infections in two ways. First, it seals up cuts or other damage to the skin by clotting (setting). This stops more blood escaping and prevents dirt and germs from entering. Second, special white blood cells called lymphocytes make antibodies, which recognize germs and attach to them. The germs can then be destroyed by other defences, including white blood cells called phagocytes that eat them.

• A drop of blood contains about 100 million red blood cells. More than 2 million of them are destroyed and replaced every second.

• The English doctor William Harvey (1578–1657) discovered that blood circulated round the body. Before that doctors had believed that our blood was continually destroyed, and that the food we ate turned into new blood to replace it.

find out more
Hearts
Lungs

Bones

Bones are the hard parts of your body. They form your skeleton, the frame around which your body is built. They give you your shape and enable you to move. Bones contain living cells and can regrow if they break. They are as strong as some kinds of steel, but only one-fifth as heavy.

Bones make up the skeletons of all mammals, birds, reptiles, amphibians and most fishes.

Bones are made out of a mixture of a tough protein called *collagen*, and very tiny, hard mineral crystals which contain calcium and phosphorus. This mixture gives bones their strength. At the centre of many bones is a soft tissue called *bone marrow*. This

contains blood vessels that supply the bone with food and oxygen. The marrow is also the place where new blood cells continually are made.

As a young mammal grows in its mother's womb, its bones develop. To begin with they are made of a softer, gristly substance called *cartilage*. During growth most of the cartilage becomes bone, apart from those parts that remain bendy, like the tip of the nose and the ears. In an adult cartilage also forms the smooth surfaces where bones slide against one another at joints. In some fishes, such as sharks and rays, the cartilage never develops into bone.

Some animals have unusual bones. Deer's antlers are bones. They grow from the skull and are shed each year.

shaft

marrow

this end fits into hip bone

◀ A human thigh bone, cut away to show the internal structure.

compact bone

spongy bone

this end forms part of knee joint

find out more
Muscles
Skeletons

Brains

Most animals have a brain, which controls their thoughts and actions. The brain consists of many connected nerve cells (neurons). Some of these pass information from sense organs, such as the ears or eyes, into the brain, while others are connected to nerves that lead from the brain to muscles.

find out more
Animal behaviour
Cells
Nervous systems
Senses

A mammal's brain is the central part of a highly complex nervous system. Encased within the skull, which protects it, the brain is connected to nerves in the spinal cord. The cord runs down from the head inside the backbone. Nerves from the spinal cord are connected to muscles, while other nerves from sense cells in the skin and muscles connect back into the spinal cord.

Processing and control

Most of the brain cells in mammals are connected to other brain cells and process incoming information, carry out thought processes and make complex decisions. Even smaller and less intelligent animals, such as bees, can remember where their hive is and calculate the time of day.

Sense organs rapidly pass information to different parts of the brain as a series of nerve impulses, which act as signals. These may be simple signals, giving information about what part of your body has been touched, or a very complex series of signals using thousands of nerve cells to allow you, for example, to see the shapes of the letters and read the words on this page.

Using information from different parts of the brain, an animal sends signals to its muscles so that they move in a particular way. Some types of movement, such as a single kick, do not require much control, but walking or flying require exact control of the muscles. You would fall over and bump into things if you could not adjust your muscles continually.

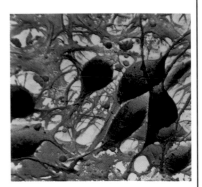

▲ Brains consist of millions of nerve cells like these (shown here magnified many times), carrying millions of messages to and from the brain. All nerve cells have the same basic structure: a large cell body, with a central nerve fibre connecting the cell to thousands of others.

Thinking, memory and learning

In humans the cerebral hemispheres, the main sections of the brain, are very large and are involved in consciousness, thought, recognition, memory and personality.

Memory is the ability to recall things and feelings from the past. Part of the action of storing a memory seems to be the making of new links between nerve cells in the brain. It is easier to remember things if you repeat them many times, or if they are very unusual or unpleasant.

Most animals with brains are able to learn. Animals with big, complicated brains are usually able to learn more than those with small brains. Having a large memory, gained through life's experiences, allows an animal to make more complex decisions and generally to respond in a more intelligent manner. Humans, with their large brains and powerful memories, are probably the most intelligent animals.

1

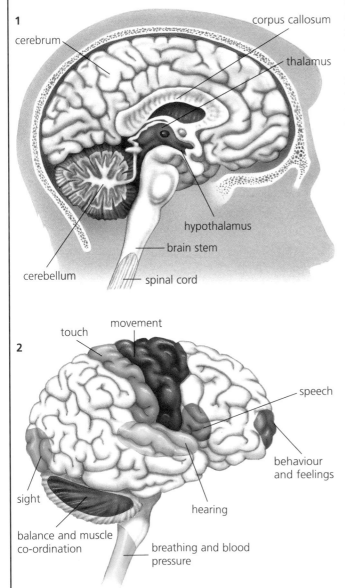

cerebrum
corpus callosum
thalamus
hypothalamus
brain stem
cerebellum
spinal cord

2

touch
movement
speech
behaviour and feelings
hearing
sight
balance and muscle co-ordination
breathing and blood pressure

◀ A cross-section through the centre of the human brain (1). The three main parts of the the brain are the brain stem, the cerebellum and the cerebrum. The cerebrum is divided into two halves, the left and right cerebral hemisphere, separated by a deep groove. Different parts of the brain carry out different jobs in the body (2). The cerebrum controls the senses.

Breathing

Breathing is the forcing of air in and out of the lungs. The body gets the energy it needs by a process called respiration. This process is fuelled by the gas oxygen, which is extracted from air in our lungs.

When you breathe in (inhale), your chest gets bigger as air is sucked into the lungs. This happens because muscles pull the ribs upwards and outwards, and a sheet of muscle below the lungs, called the diaphragm, is pulled downwards. When you breathe out (exhale), your ribs get lower and are drawn in and the diaphragm rises.

The brain controls the speed of our breathing. This is to provide enough oxygen at all times for the body's needs. When you are still, your energy needs are low; you require little oxygen and your breathing is slow and shallow. You will probably breathe in and out about 12 times every minute. However, if you start exercising hard, you need much more energy. Without thinking about it, you automatically find yourself breathing faster and more deeply.

Respiration

Respiration is the process by which the oxygen we breathe in reacts with food to release energy. The oxygen passes into the red blood cells through tiny blood vessels in the lungs. The blood then carries the oxygen to every cell in the body where respiration occurs. In these cells oxygen is combined with sugars and fats in food to release energy plus some water and the waste gas carbon dioxide. The blood then carries the carbon dioxide back to the lungs, from where it is breathed out every time we exhale.

Some creatures can get their energy without oxygen. This is called anaerobic respiration. They include microbes living in thick mud, and parasites living inside an animal's gut. When you run very quickly, your muscles respire anaerobically because they cannot get enough energy from normal respiration. This produces lactic acid as a waste product instead of carbon dioxide, and it is this acid that causes your muscles to ache.

- People usually think that respiration means breathing. However, breathing is just the means of getting oxygen so that respiration can take place.

- Some animals do not use lungs and air for respiration. For example, fishes use their gills to get oxygen from the water.

- Like animals, plants need oxygen for respiration. However, they use less oxygen in respiration than they create in the process of photosynthesis. In this process they use energy from the Sun to turn carbon dioxide in the air and water into glucose and oxygen.

▶ Breathing in and breathing out. All mammals, birds, reptiles and air-breathing amphibians breathe in this way.

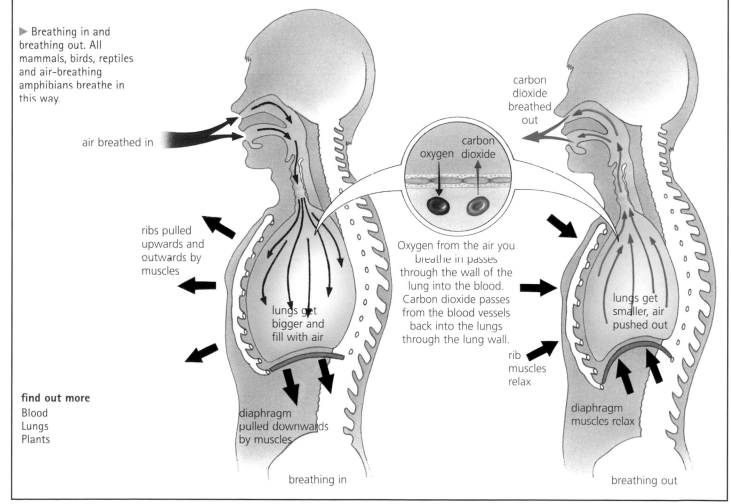

air breathed in

ribs pulled upwards and outwards by muscles

lungs get bigger and fill with air

diaphragm pulled downwards by muscles

breathing in

oxygen · carbon dioxide

Oxygen from the air you breathe in passes through the wall of the lung into the blood. Carbon dioxide passes from the blood vessels back into the lungs through the lung wall.

carbon dioxide breathed out

lungs get smaller, air pushed out

rib muscles relax

diaphragm muscles relax

breathing out

find out more
Blood
Lungs
Plants

Butterflies and moths

Butterflies and moths are familiar insects which are found almost everywhere in the world. Most butterflies are brightly coloured, day-flying insects, while most moths are active at night and are drab in colour.

Most butterflies and moths live for less than a year. They cannot chew their food. Instead, they use their long, tube-like tongues to suck up liquids, such as the sugary nectar from flowers.

The surface of a butterfly's wings is covered in thousands of tiny scales and is brightly coloured. The underside is often grey and brown. When its wings are folded, it is usually hidden from creatures that might eat it. Moths too are difficult to see when at rest.

Because of their beauty, butterflies have been collected on a vast scale. Some of the big tropical butterflies, such as the morphos from Brazil, have been used as jewellery. Even some of the small European butterflies are in danger of dying out, largely because of the destruction of their habitats and the plants on which their caterpillars feed.

Mating, eggs and caterpillars

Butterflies find and court their mates with a show of colour. Moths use their excellent sense of smell to find a mate. After mating, female butterflies and moths lay batches of eggs on suitable food plants, and then die.

These eggs hatch into caterpillars, which are similar to the maggots or grubs of other insects. Caterpillars spend most of the time feeding. When they have grown big enough, they pupate (enter a resting stage). In the pupa (chrysalis) the butterfly develops and then emerges as an adult.

◀ This monarch butterfly has just emerged from its pupa, and is spreading its wings for the first time.

• There are about 20,000 different kinds of butterfly and about 120,000 different kinds of moth.

• Many butterflies travel long distances on migration. Comma butterflies have flown from the Sahara to northern Europe, a distance of 3000 km, in 14 days.

find out more
Flowering plants
Insects
Migration

Cactuses

Most cactuses are spiky plants that grow in very dry, hot areas of North and South America. There are many hundreds of kinds of cactus, ranging from tiny plants that look like pebbles to giants as tall as trees.

Cactuses (or 'cacti') are well suited to living in dry, desert conditions. They have very long roots, which grow close to the surface and enable them to take up water quickly during rare storms. They store the water in their thick, tough stems.

The spines of a cactus are actually its leaves and, because they are so tiny, they lose very little water through evaporation. The spines also stop animals from trying to eat the cactus.

Most cactuses grow very slowly. Some only grow about 2 centimetres in a year. Like other plants, cactuses use sunlight to make food, by the process called *photosynthesis*. In most plants photosynthesis takes place in the leaves, but since the leaves of a cactus are so small, they photosynthesize their food in their stems instead.

Most cactuses have colourful or unusual flowers, and this makes them popular with gardeners, who also like their unusual shapes. Some cactuses, such as the prickly pear, are edible. Others, such as the barrel cactus, can be used as a source of water by people in the desert. Some Native Americans have a tradition of eating one kind of cactus, the peyote cactus, to bring about dream-like states on special religious occasions.

find out more
Deserts
Flowering plants
Plants

▶ The saguaro cactuses of Mexico and the southwestern United States grow up to 15 or 20 m high, and may live for 200 years. They often do not flower until they are 50 or 75 years old.

▶ The golden barrel cactus has such strong spines that people have used them as tooth picks. This cactus can grow up to 90 cm wide.

Camels

Camels are the largest animals of the desert. They are well suited to the harsh, waterless environment. For centuries they have been domesticated and used to carry heavy loads long distances across hostile landscapes.

Camels' bodies are adapted to save water in the dry desert.

They do not sweat, and they lose very little moisture in other ways. But most importantly, a camel can go without drinking for long periods, losing up to 40 per cent of its body weight. A camel will replace the liquid it has lost as soon as it reaches water and may drink over 50 litres in one go. A camel's two-toed, broad-padded feet are ideal for walking on loose sand or gravel, and it can close its nostrils against sandstorms.

There are two kinds of camel, and both have been domesticated. The one-humped Arabian camel, from the Middle East, is no longer found in the wild. However, some domesticated ones have been released in parts of the world (such as Africa and Australia) and have become semi-wild. The two-humped Bactrian camel, from the rocky deserts of central Asia, is now rare in the wild.

The llama and its close relative the alpaca are grazing animals from South America. They are related to camels, although they have no hump. They are valued for their long, fine wool. Llamas are also used to carry heavy loads in high mountain areas.

◀ The Bactrian camel has quite long, shaggy fur which helps keep it warm in the extreme conditions in which it lives in the wild.

• There are six members of the camel family: the Arabian camel, the Bactrian camel, the llama, the alpaca, the guanaco, and the vicuña.

• Camels' humps do not contain water but fat, which is used up in times of hardship.

◀ The Arabian camel (or dromedary) has longer legs and is not as heavily built as the Bactrian camel.

find out more
Deserts
Mammals

Camouflage

Most kinds of wild animal are hard to see because they are camouflaged – coloured or shaped to match the places in which they live. Their camouflage is a source of protection because they cannot easily be seen by animals that hunt them.

Hunters in turn need camouflage, both for protection against animals that hunt them and so that they remain unseen as they close in on their prey.

Countershading, where an animal's back is darker than its underside, is a form of camouflage that almost every animal has. Stronger light falling on the animal's back makes its rounded body appear to be all one colour and flat. It is then harder to see.

Some animals have *disruptive coloration* – their skin or fur is broken up by spots or stripes. This helps them merge into the dappled light around them. They may also have a *disruptive shape* – the outline of their body is broken up with spines or flaps of skin, making them difficult to see. These bits may be shaped and coloured to look like leaves or thorns on a plant.

Some camouflage copies the patterns and colours of animals that taste unpleasant or have a sting. For instance, harmless hoverflies look like wasps, which scares away hunters such as dragonflies.

Although camouflage is vitally important, it is not enough. Stillness and silence are part of the trick of not being seen.

find out more
Animal behaviour
Food chains and webs
Lizards

◀ Some of the most effective camouflage is found in tropical rainforests. This leaf-imitating insect in the Costa Rican rainforest is very hard to tell apart from the real thing.

Cats

The cat family includes wild cats such as lions, tigers, leopards and cheetahs, and domestic cats that people keep in their homes as pets.

serval

Cats are supreme hunters. Their diet consists mainly of meat, so wild cats must kill to survive. Most cats are very fast over a short distance, but they give up if their prey runs too far. They have lithe bodies and strong legs with padded feet that enable them to move silently. Their eyesight is also good, and many of them hunt at night or when the light is dim. Their eyes are large and look forward, rather than seeing all around.

Cats have an excellent sense of hearing, and the long whiskers growing around their face also help them to sense their surroundings. If their whiskers are touched, even very lightly, cats are aware of it. This not only enables them to avoid bumping into things at night, but they may be able to feel the disturbance of the air if their prey makes a dash for safety.

Most cats use their strong and sharp claws to catch their prey. All cats except cheetahs can retract (pull in) their claws, so they do not become worn and blunt when they walk.

• There are 37 kinds of wild cat. The lion, tiger, jaguar, leopard and snow leopard are known as 'big cats'. Smaller wild cats include the lynx from Europe, Asia and North America, the ocelot from Central and South America, and the serval from Africa.

▼ Tigers are protected animals but they are often killed illegally, even in nature reserves. Unless this killing is stopped, it is likely that there will be no tigers left in a few years time.

▲ A lion eating a zebra that it has killed. An adult male lion may eat 40 kilograms of meat in a single meal, but then he will probably not feed again for several days.

Lions

Lions live in grassland or scrub country. A few hundred remain in one reserve in India; the rest live in Africa. Unlike most members of the cat family, lions live in groups, called prides. A pride usually consists of about a dozen animals. At least two of them are adult males, the rest females and their cubs. The females in a pride are usually sisters, and remain together all their lives. The males rarely stay with the same pride for more than three years.

The lionesses do most of the hunting. Sometimes two or more work together to stalk an antelope or zebra, which are their main food. However, although the females may make the kill, the males usually feed first.

Tigers

Tigers are big cats that live in forested areas. At one time they were found throughout much of eastern and southern Asia, but now they survive mainly in India, Sumatra and the far east of Siberia.

Male tigers live alone in large territories, although these may overlap the living areas of several females. They are chiefly active at twilight or at night. They may travel up to 20 kilometres in a night and can leap as far as 10 metres in a single bound. They are good swimmers and can climb quite well. Their main food is large mammals such as wild pigs, buffalo and deer.

Leopards and cheetahs

Leopards live in much of Africa and Asia, in forests, grasslands and even deserts. They usually live alone and defend their territories from others. They climb trees very well and often rest up in the branches. They also hide the remains of their meals in trees, out of reach of most scavengers. They mainly hunt medium-sized grazing animals such as small antelopes and pigs, but will resort to smaller prey, even insects.

▼ The cheetah is now a protected animal. The main threat to its existence is the destruction of its natural home and the prey it finds there.

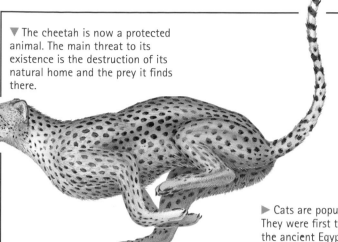

▶ Cats are popular pets. They were first tamed by the ancient Egyptians at least 3500 years ago.

Cheetahs usually live in small groups on grassy plains in eastern and southern Africa. They are able to hunt in the daytime because they can run faster than any other animal, reaching up to 100 kilometres per hour. But they cannot keep this speed up for long, and an average chase lasts for less than half a minute. They usually hunt small antelopes or the young of some larger animals. A cheetah's claws are blunt, so it kills its prey by knocking it off balance and then throttling it.

Jaguars and pumas

The jaguar's beautiful blotched coat camouflages it in the South American forests, where it is usually found. Jaguars live alone in large territories. They are active mainly at night, when they hunt large ground prey such as peccaries, capybaras and tapirs. They are good swimmers and sometimes feed on fish. They occasionally even catch caymans.

Pumas (also known as cougars or mountain lions) live in remote parts of North and South America. They live alone and do not have fixed dens; instead they travel through their territory, usually over 30 square kilometres. They have very good eyesight and hearing, but their sense of smell is poor. They have powerful hind legs, and it is said that they can leap upwards more than 5 metres. Their main prey is deer, usually sick or old ones. They make about one kill a week, and drag the carcass to a safe place to be eaten over several days.

Cats in danger

Many of the world's wild cats are now close to dying out. This is partly because of the destruction of their habitat.

Also, some cats are hunted because they have attacked domestic animals or even humans, and others are killed by hunters. Many are killed, however, for their beautiful, thick, soft fur, which is often patterned with spots or stripes. The fur is highly prized for making coats or even rugs. Today many cats are protected by law in most parts of the world.

• There are over 500 million domestic cats around the world.

find out more
Ecology
Grasslands
Mammals

▼ Pumas rarely attack domestic animals. Even so many are shot by farmers and cattle-herders.

▲ The number of jaguars in the world today is very low because of hunting and forest clearance by humans.

Record-breaking cats
Largest
Lions, shoulder height up to 1.2 m
Largest domestic breed
Ragdoll, males weigh up to 9 kg
Smallest domestic breed
Singapura, males weigh up to 2.7 kg, females up to 1.8 kg
Fastest
Cheetah, up to 100 km per hour

Cattle

Cattle include a number of large plant-eating animals such as cows, buffaloes and bison. A few kinds of cattle are still found in the wild, but most live as farm animals and are a vital source of milk, meat and leather.

There are several different kinds of wild cattle, including yaks, bison and buffaloes. Many live in forested areas where they can find plenty of food and safe, shady places to rest. However, some, such as yaks, live in high mountain areas of the Himalayas. Others, such as North American bison, live on vast plains where they graze on the plentiful grass.

Cattle are social animals and live in groups led by a bull. Both the males and females have horns. They are straight or curved and last all through the animal's life. The horns of water buffalo are bigger than those of any other animal.

Cattle are now very rare in the wild, largely as a result of human hunting, but they are common as farm animals. Most farm cattle are descended from the aurochs, which once roamed Europe but became extinct about 350 years ago. Today there are about 200 different domestic breeds. They are mostly bred for their milk, meat and leather. However, some are also used for pulling ploughs and heavy loads. Cattle droppings can be used as fertilizer and sometimes they are dried out and used as fuel on fires. Some cultures also use cattle as a measure of wealth: the more cattle a person has, the richer they are.

• Cattle digest their food using a complex, four-chambered stomach. They are known as ruminants, as are antelopes, sheep, deer and giraffes. All these animals re-chew food that has been swallowed and then brought back into the mouth. This process is known as 'chewing the cud'.

• There are 12 different kinds (species) of cattle.

find out more
Digestive systems
Mammals

▲ Domestic cattle, such as the Friesian (left), look quite different from wild cattle, such as the gaur of South-east Asia (right).

Cave animals

In the past caves provided shelter for a number of large animals, such as the giant cave bear, which have now died out. In prehistoric times caves also provided a safe home for many humans.

Most animals that live in caves today are quite small. Deep inside a cave it is completely dark, and the temperature is the same all the year round. The few animals that live all their lives in the dark are often white or transparent. Most of them are also blind, but they have a very good sense of touch. Cave animals include many special kinds of fishes, shrimps, salamanders, centipedes, spiders, scorpions and insects.

No plants grow inside caves because plants cannot grow without light. Cave animals have to find other sources of food. Animals such as bats and birds, which use the cave as a shelter, fly out to feed outside the cave. Their droppings and dead bodies provide food for many small cave creatures, which in turn are eaten by larger animals.

▼ The olm, a kind of cave salamander, is born with good eyes and some colour. By the time it is 18 months old it is totally white and completely blind.

find out more
Amphibians
Bats
Centipedes and
  millipedes
Crabs and other
  crustaceans
Ecology
Insects
Newts and salamanders
Spiders and scorpions

Cells

Cells are the building blocks of life. Some very simple plants and animals have only one cell, but most living things are made of huge numbers of cells. A newborn baby has about 5 million million cells in its body, and an adult has over 10 times that number. Most cells are so tiny that they can be seen only by using a powerful microscope.

Cells can look different, but most have the same basic parts. They are each surrounded by a cell membrane which holds them together and controls which substances can enter or leave. Inside this membrane, the cell is divided into two parts: the nucleus and the cytoplasm.

The *nucleus* contains the body's genes. Genes control what features a plant or animal will have. They contain a chemical, DNA, which forms a kind of code or blueprint, setting out how the cell will develop. Genes are arranged in groups on structures called *chromosomes*. When a cell divides, each chromosome makes an exact copy of itself so that the new cell has a complete blueprint.

The *cytoplasm* surrounds the nucleus and contains food and many other substances. It is a kind of chemical factory where the materials needed for living, growing and changing are produced. The nucleus controls what goes on in the cytoplasm by sending chemical messages to it. The cytoplasm contains a number of structures called *organelles*, each surrounded by its own membrane.

Different organelles have different tasks. The *endoplasmic reticulum* makes various chemicals that the body needs and breaks down waste materials. The *mitochondria* are the powerhouses of the cell. Here, the cell uses oxygen to break down glucose from food to provide energy. In plant cells, organelles called *chloroplasts* contain the green pigment chlorophyll, which traps the energy of sunlight. This is used to build up sugars for food from carbon dioxide and water.

Not all cells contain a nucleus. In very simple cells, such as those of bacteria, the DNA is free in the cytoplasm. Instead of chromosomes, there is a single circular strand of DNA. These cells do not have organelles either.

Tissues and organs

Tissues are groups of similar cells that work together to perform a particular function. For example, muscle tissue is made up of cells called muscle fibres. These contain protein fibres which can shorten (contract) in response to nerve signals. Nerve tissue consists of cells with long, thin nerve fibres and a cell membrane that can transmit electrical nerve impulses.

Organs are groups of different kinds of tissue, containing different types of cell, which work together to carry out a particular function. For example, the heart contains muscle tissue for pumping, nerve tissue for controlling the pumping, and blood vessels for supplying the other cells with food and oxygen.

• The oldest forms of life are microscopic, single-celled organisms, such as bacteria. Over time, groups of single-celled organisms began to cluster together in colonies. Some simple animals, such as sponges, are in some ways like masses of colonial cells living in a single co-operative structure.

find out more
Animals
Bacteria
Genetics
Living things
Plants
Viruses

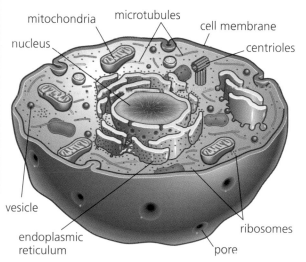

▲ A typical animal cell. Animal cells have a faint outline when seen under a microscope, because they are enclosed only in a thin cell membrane.

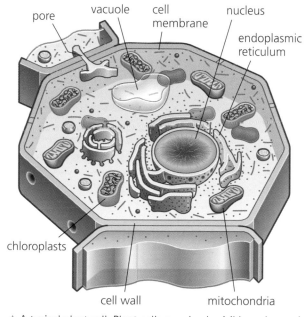

▲ A typical plant cell. Plant cells are clearly visible under a microscope because they have a thick cellulose cell wall outside the cell membrane.

Centipedes and millipedes

Centipedes and millipedes are best known for the extraordinary number of legs they have. As a group, they are called myriapods, which means 'many-footed animals'.

The legs of these animals are attached to a series of segments that make up the body. Centipedes have one pair of legs on each segment, but millipedes

• There are over 2000 kinds of centipede and about 8000 kinds of millipede.

find out more
Animals
Skeletons

▼ The centipede's flattened body enables it to squeeze through narrow spaces in soil or rotten wood. Its eyesight is poor, but its long antennae help it to track its prey.

one segment

eyes

antenna   fang

last walking leg

▶ Millipedes have a row of stink glands along their sides, from which they produce unpleasant chemicals when disturbed. Some millipedes protect themselves by rolling up into a tight ball.

have two. Like insects and spiders, they have external skeletons.

The name centipede means 'hundred feet'. A few kinds of centipede have far more than this, but most have fewer. At first, young centipedes do not have many feet, but they grow more legs as they grow larger and longer. Centipedes move fast, for they are hunters, tracking insects, grubs and worms. They use poison fangs to overcome their prey. Some tropical centipedes can give an unpleasant bite if they are handled, but most are entirely harmless to humans.

one segment   antenna
eye

mouthparts

opening of stink gland

two pairs of legs on each segment

The name millipede means 'thousand feet', but none of them has more than 400 feet. They cannot move fast, but their legs provide enough power to burrow through leaf litter and loose soil to search for soft or decaying plants to eat. Those in forests are good recyclers, returning chemicals to the soil for plants to use again.

Cereals

find out more
Grasses

One of the most important kinds of food we eat is made from the seed of a type of grass. Since prehistoric times people have grown plants of the grass family, called cereals. The most common cereals are rice, wheat, maize (corn), oats, rye, barley and millet.

The seeds of cereal plants are called grains. They are quite large and full of carbohydrates, which are filling and provide the body with a quick source of energy. The grains also contain some

protein, vitamins and minerals, and have plenty of fibre. Most of the fibre is found in the bran, the grain's outer layer.

Cereal grains can be eaten whole or processed. Ready-to-eat cereals are made mainly from maize, rice and wheat. Some breakfast cereals use the whole grain and they are more nutritious than those that do not. Cereal grains can also be ground into flour to make bread, pasta, porridge and puddings.

rice

barley

millet

oats

rye

wheat

maize (corn)

Conservation

With more and more people wanting space for houses, factories and farms, there is less and less space for other living things. Conservation is action to stop rare animals and plants from dying out and to stop common animals and plants from becoming rare. It also aims to protect the wild places where all these animals and plants live.

In the past, some animals became extinct (died out) because they could not cope with natural changes around them. The dinosaurs may all have died out when a meteor from space crashed into the Earth. This sent clouds of dust into the air, which blotted out the Sun and made the weather too cold for them. However, humans have greatly speeded up the rate at which plants and animals are becoming extinct.

For example, the dodo that lived on the island of Mauritius was a slow-moving bird that could not fly. When the first humans discovered Mauritius, they found that the dodo was easy to kill and good to eat. In less than a century, they had killed every dodo. Many other animals have also been made extinct by humans killing them and destroying their homes.

Animals in danger

Many other living things are in danger of becoming extinct. At least 1000 different kinds of birds and mammals are now extremely rare, and nobody can even begin to guess how many kinds

▲ No more than 400 mountain gorillas survive in the mountain forests of central Africa, yet some are still shot by poachers. A recent war in the countries where they live has reduced numbers still further.

of insects and plants are in danger. The giant panda from China is a famous example. Less than 1000 pandas survive in the wild, because so much of the wild bamboo forest in which they live has been chopped down by people to make

▼ For many centuries, animals have been made extinct by human hunting. In the last 350 years, 95 different birds and 40 different kinds of mammal have become extinct, including the penguin-like great auk from the islands of the North Atlantic, the passenger pigeon of North America, and the quagga (a kind of zebra) from Africa.

• The rarest animal in the world is probably the Spix's macaw. A single male bird was found in a wood in Brazil in 1990. About 30 other Spix's macaws live in zoos, and a few females have now been released back into the wild, in the hope that they will breed with the male to save the species.

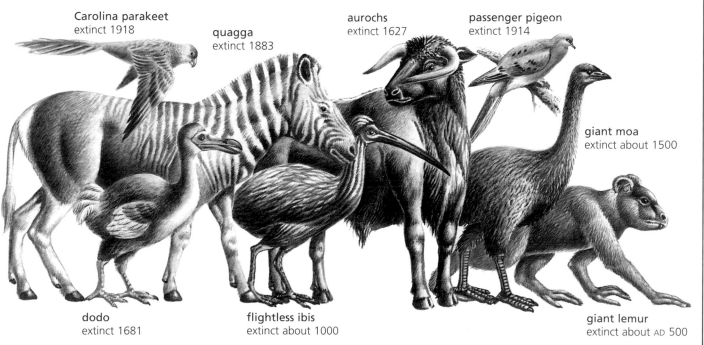

Carolina parakeet
extinct 1918

quagga
extinct 1883

aurochs
extinct 1627

passenger pigeon
extinct 1914

giant moa
extinct about 1500

dodo
extinct 1681

flightless ibis
extinct about 1000

giant lemur
extinct about AD 500

way for fields. Several kinds of rhinoceros are in danger because of hunting for their horns, and tigers are threatened by hunting for their skins.

Rare birds that need conservation include the bald eagle – the national bird of the United States – and the Mauritius kestrel from the island where the dodo once lived. Thanks to the work of conservationists, these animals are now commoner than they were, but they still need help if they are to survive.

None of these plants and animals can survive unless we protect their habitats – the wild places in which they live. Many of the rarest animals live in the tropical rainforest, but the rainforest itself is in danger. Every second, an area of rainforest the size of a football pitch is cut or burnt down, either for the valuable timber from its trees or to make way for farming. All the animals that live there then lose their homes.

In the sea, coral reefs are almost as full of life as rainforests, but only about a tenth of them are still in good condition. The rest have been damaged by waste chemicals, by mud washed down rivers, by too many human visitors and even by coral mining.

Positive action

Since about 1970, people have realized how seriously these plants, animals and wild places are threatened, and have begun to do something. Conservation charities like the World Wide Fund for Nature (WWF) have spent huge sums of money trying to protect rare animals and plants and their habitats. Governments have passed laws to stop animals being killed, and they have set up nature reserves and national parks to keep their best pieces of countryside safe.

Perhaps the most important change is that people have begun to recognize the value of wild plants and animals. As well as the pleasure we

find out more
Ecology
Pollution
Zoos

▼ People who protect wildlife and wild places are called conservationists. These young conservationists are cleaning up a river to make it a better home for plants and animals.

▲ Every year, an area of rainforest twice the size of Austria is destroyed by humans. In this satellite photograph, taken over western Brazil, the natural forest (shown in pink) is being cut down to make way for farmers' fields (shown in blue and green).

get from seeing a beautiful animal in its habitat (even if it is only on television), many living things have other values for humans. For example, a medicine from a rare plant called the rosy periwinkle from the rainforest of Madagascar is now used to cure a disease called leukaemia. The rainforest and the oceans also have an important role in locking up the gases, produced by factories and power stations, that are causing global warming. However, this can only happen if these wild places are kept in good condition.

The variety of life

In 1985 an American scientist invented a new word: *biodiversity*. It combines the words 'biological' (referring to living things) and 'diversity' (meaning lots of different kinds), so it means the total variety of all living things on Earth. Biodiversity is part of the natural 'wealth' of our planet. We all become a little poorer for each bit of biodiversity that is lost. The sheer variety of living things also helps to keep the environment in good working order. Biodiversity is therefore an important measure of the health of the planet.

In 1992, at an international meeting called the Earth Summit in Rio de Janeiro, Brazil, world leaders signed a document promising they would protect the biodiversity of their countries, and help poorer countries do the same. That means that everyone around the world has a part to play in saving living things and their habitats.

Crabs and other crustaceans

Crustaceans are a large and successful group of animals including crabs, lobsters, barnacles and woodlice. The name crustacean means 'crusty one', a suitable name for animals that are covered in hard armour.

Like their relatives the insects, crustaceans have jointed legs; unlike insects, their head and thorax (the middle part, which is separate in an insect) are usually joined into one segment, and they have two pairs of antennae (feelers), not one. Most crustaceans spend their lives in the sea or in fresh water.

• There are about 150,000 different kinds of crustacean. Of these, over 6000 are lobsters and about 4500 are crabs.

▼ Members of the main crustacean groups. With the exception of the shore crab, all these animals are shown five times larger than they are in real life.

▲ Lobsters, such as the common lobster shown here, belong to the 'walking' group of crustaceans. Large lobsters have very big 'claws' or pincers on their first pair of legs. One of the claws is much larger than the other and is specially designed for stabbing or crushing prey.

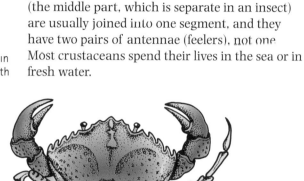

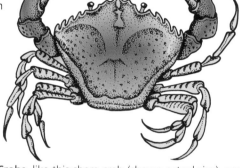

Crabs, like this shore crab (shown actual size), are heavily armoured and most live in the sea. Many are fast-moving scavengers. The biggest crab is the Japanese spider crab, which can measure up to 2.5 m across with its legs stretched out.

Daphnia, or water fleas, seem to jump through the water as they swim. *Daphnia* females produce young without mating.

Fish lice are parasites that feed on the blood of their host. They often live in the gills of fish.

Acorn barnacles are protected by plates of shell. Their hairy feet comb the water for fragments of food and for oxygen.

Cyclops swim jerkily, like *Daphnia*. Most live in still, fresh water. The females carry two egg sacs.

Ostracods are sometimes called seed shrimps. Their hinged shell makes them look like small seeds in the water.

Some of the most ancient fossils known are crustaceans, dating back over 500 million years. Most of these early types had a large number of legs which, as they swept through the water, gathered fragments of food from it.

Today, many small, swimming crustaceans are filter-feeders – the legs have fine hair-like projections which sweep food in the water towards the animal's mouth. Other crustaceans live on the sea-bed and have legs suitable for walking and grasping food.

Crustacean variety

The appearance and size of crustaceans are extremely varied. Lobsters are the giants among crustaceans, the biggest weighing over 20 kilograms. The smallest are tiny copepods, just a fraction of a millimetre long, which live between grains of sand on beaches or float in the sea.

Many crustaceans, such as shrimps, prawns and lobsters, have long bodies with a fan-like tail at the end. Some, like ostracods, have a hinged shell that makes them look like small seeds.

Life cycles

Crustaceans tend to be long-lived animals. Although some kinds may migrate short distances to suitable breeding places, in general, once they are adult, they live in a small area and do not move from it. Only in the earliest phases of their life after hatching do they travel freely.

Most marine (sea-dwelling) crustaceans produce large numbers of tiny larvae – young forms that are very unlike their parents. At this stage, they are part of a mass of plankton (minute plants and animals), swept along by the currents of the oceans. Many of them are eaten by fishes or other animals.

As they grow, crustaceans cast off and regrow their shell many times. During the course of its life a large crab or lobster may shed its shell 20 times. Each time, it emerges with a new, soft covering and must remain in hiding until this has hardened into a strong armour. The larger the animal grows, the longer this takes.

Crabs

Crabs are amongst the most familiar crustaceans. They are mainly scavengers, feeding on dead and rotting remains, and small-scale hunters. In turn they are preyed on by many creatures, including humans, who prize the flesh of these large crustaceans.

Some crabs camouflage themselves with sponges or pieces of seaweed that they attach to their shells. Hermit crabs have long, soft bodies which they hide inside the empty shells of sea snails. Some kinds then place sea anemones on their shell. They probably do this to frighten off any hunting animals, as the sea anemone has a painful sting.

Krill and shrimps

Shrimps and various shrimp-like crustaceans are extremely abundant in the oceans. Many creatures that look like shrimps live in the open sea, though they may not be closely related to the inshore kinds. Krill are probably the most important of these. Huge numbers live in the cold waters of the Arctic and Antarctic seas, where they feed on microscopic plants and animals. In turn, krill are the basic food for many birds, seals and baleen whales. Like many other

▲ The animal plankton shown here are the larvae (young) of various different kinds of crustacean, including krill. In the Atlantic Ocean, between January and April, there are as many as 20 kg of krill in every cubic metre of water.

marine crustaceans, they have a row of light-producing organs down each side.

Living in tropical seas is a shrimp that uses sound as a weapon. Pistol shrimps have a large claw that can make a sharp, cracking noise. At close range small prey can be stunned or even killed by the noise.

Other shrimps often seen in warm waters are 'cleaners', or barber shrimps. They help fishes by pulling off and eating tiny parasitic animals living on their skin and in their gills. In return the shrimps are not harmed by the fishes, which in other circumstances might eat them.

Copepods and barnacles

There are many different kinds of copepod, which are also amongst the most abundant animals in the oceans. They and their relatives have a very important position at the base of ocean food chains. They feed on the smallest sea-living plants, and they themselves are eaten by other invertebrates and fishes. These are in turn eaten by bigger hunters.

Goose barnacles are among the animals that feed on copepods. Unlike the acorn barnacles of the seashore, goose barnacles have fragile shells. But they feed in the same way as shore barnacles, kicking long, feathery legs into the water to comb tiny sea creatures such as copepods into their mouths.

▲ The delicate and dainty arrow crab lives in the Atlantic around the coasts of North and South America. To feed, it perches on underwater rocks and gathers detritus (bits that are floating around). Later, it picks out edible particles from the detritus.

• The commonest land crustaceans are woodlice. They breathe with gills, which must be kept moist, so they cannot survive in really dry places. They mainly feed on rotting vegetation.

• Shrimps and prawns are important as human food. They are caught in huge numbers from inshore waters around the world. In recent years many prawn farms have been set up, often causing the destruction of valuable mangrove swamps.

find out more
Animals
Food chains and webs
Oceans and seas
Seashore
Whales and dolphins

Crocodiles and alligators

Crocodiles and alligators are hunting animals that live in rivers, lakes and the edge of the sea in warm parts of the world. They belong to an ancient group of reptiles which has changed little since the days of the dinosaurs.

Crocodiles and alligators are ferocious hunters when in water, using the power of their flattened, oar-like tails to drive them towards their prey. Their jaws are set with sharp pointed teeth, which are ideal for holding fish, their main source of food. Big, old crocodiles can tackle larger creatures and can sometimes be a danger to humans and livestock.

A female usually makes a nest of mud and plant debris near the water's edge, although some kinds dig a hole in the sand. She lays her eggs and stays on guard until they are ready to hatch. When they hatch, she gathers them into her huge mouth and carries them to the water. She continues to guard them as they may be eaten by other animals.

Large crocodiles and alligators have few natural enemies, as they are protected by a thick, leathery skin armoured with small plates of bone. However, the skin of young ones makes valuable leather, and many have been killed for this. Some are farmed for their leather and meat, but many kinds of crocodile in particular are endangered and need careful conservation.

- Crocodiles may grow up to 6.5 m long; alligators up to 5.8 m.

- There are 13 different kinds of crocodile, 2 alligators, 5 caymans (similar to alligators), and 1 gavial, which has a long, thin snout.

find out more
Reptiles

▼ Alligators have shorter, blunter snouts than crocodiles. No lower teeth can be seen when an alligator shuts its mouth. However, when a crocodile's mouth is closed, the large tooth fourth from the centre on the lower jaw still shows near the front of the mouth.

Chinese alligator

African crocodile

fourth lower tooth visible when jaws closed

Deer

Deer are graceful animals, with slim bodies and long, slender legs. Most live in herds in cool woodlands. Deer are alert, swift-footed animals, and they use their speed to escape the many hunting animals that try to catch them.

Most deer produce only one baby at a time. The mother leaves it hidden in the undergrowth, where it is well camouflaged by its spotted coat. She returns to feed it twice a day. After a few weeks it is strong enough to join the herd.

Males of most kinds of deer have antlers. The size of the antlers shows a male's age and position in the herd. They are often used in fighting for females. Fighting males lock antlers and push, until the weaker deer breaks away. Unlike all other kinds of deer, reindeer (caribou) females have antlers, but they are smaller than those of the males.

In parts of northern Europe and Asia reindeer have been domesticated, and in a few places they are still ridden or used to pull sledges. A few kinds of deer, such as red deer, are now being farmed for their meat.

- There are 45 different kinds of deer.

- Antlers, which are made of bone, are shed and regrown each year. When they are growing, they are covered with live furry skin ('velvet') which is later rubbed off. This leaves an antler of bare bone that lasts for just one breeding season before falling off.

find out more
Forests
Mammals
Tundra

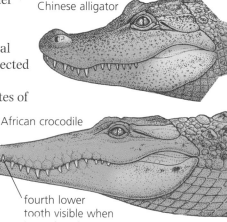

▶ The Chinese water deer is a tiny, shy creature that lives in the marshes of China and Korea.

▶ The adult male red deer is one of the largest mammals in Europe. Its antlers can grow to be more than 1 m long.

▶ The moose is the largest of the deer family. Called a moose in North America, in Europe it is known as an elk.

Desert wildlife

A desert is a large, extremely dry area of land. Geographers say that a 'true' desert receives less than 250 millimetres of rain in an average year. Yet rainfall in the desert can vary a great deal from year to year, and is difficult to predict. A desert may have heavy rains in one year, followed by several years with no rain at all. To survive in the desert, the people and wildlife living there have to adapt to the changing conditions.

- In coastal deserts, such as the Namib Desert in south-western Africa, fog may be the most reliable source of water for many smaller animals and plants.

Most deserts are in or near the tropical regions of the world. These are the *hot deserts*, where the Sun shines constantly from cloudless skies. During the night, the clear skies allow heat to escape, and it can become surprisingly cold. The deserts of central Asia are not near the tropics. They are far inland, where the winds are dry. Deserts such as the Gobi are hot in summer and bitterly cold in winter.

▲ A gemsbok (oryx) stands before sand dunes in Africa's Namib Desert. During a long period of drought, large animals such as this often move to other areas where there is more rain.

Plants and animals

- Places in the desert where water lies at or close to the surface are called *oases*. The soil here may be quite fertile, and trees, shrubs and crops can all flourish because their roots can reach water.

Desert plants and animals face great challenges in order to survive in the harsh hot desert conditions. Many animals avoid the heat by sleeping during the day and coming out to find food at night. Insects and other small animals shelter in the sand, while larger animals keep cool by seeking shade under a tree or a rockface. Many small mammals, such as jackrabbits, have very big ears. These help them to lose heat.

Desert wildlife also has to survive with little or no water for long periods of time. Camels can live for several weeks without a drink. They store fat in their humps, and their bodies produce water by breaking down this stored fat. Many desert plants store water in their stems, leaves or roots. Some have a thick waxy coating that reduces the amount of water lost through their leaves. Other desert plants have seeds that can wait many years for rain. When it does rain they quickly grow, flower, form new seeds, and die.

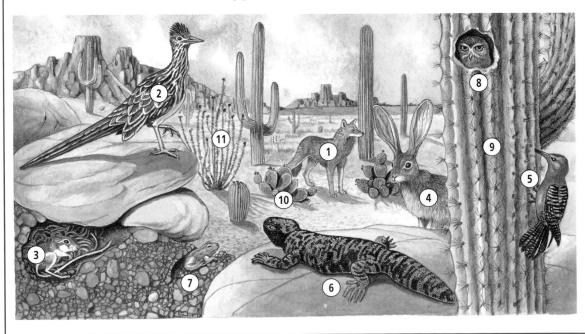

◀ Wildlife in the Sonoran Desert in Arizona, USA.
1 coyote
2 roadrunner
3 kangaroo rat
4 jackrabbit
5 Gila woodpecker
6 Gila monster
7 spadefoot toad
8 elf owl
9 saguaro cactus
10 jumping cholla
11 ocotillo

find out more
Cactuses
Camels
Plants
Tundra

Digestive systems

Your digestive system breaks down the food you eat into small particles that can be absorbed into the blood. In humans and most other animals digestion takes place in the gut, which is really a long tube that stretches from the mouth to the anus.

Most vertebrates (animals with a backbone) have teeth for biting and chewing their food to break it into small pieces which are easy to swallow. Many invertebrates (animals without a backbone) have hard mouthparts to help them break up their food. In mammals, including humans, the front teeth bite off a chunk, then the tongue pushes it to the back teeth, which grind it up and mix it with saliva. *Saliva* is produced by the salivary glands. It softens the

• The appendix is a small, worm-shaped organ in the lower gut. Most mammals have a similar organ. In plant-eaters it is used to digest grass. In humans it no longer has any function.

▶ The human digestive system is largely similar to that of other vertebrates (animals with a backbone).

▼ A carpet of villi 0.5–1.0 mm high lines the small intestine. These villi give the intestine a huge surface area for absorbing digested food. Your small intestine has an inside surface area of 10 square metres.

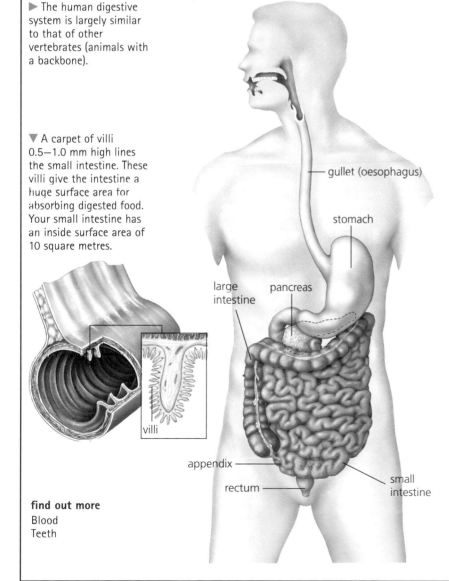

gullet (oesophagus)

stomach

large intestine

pancreas

villi

appendix

rectum

small intestine

find out more
Blood
Teeth

Rumination

Bison, like all cattle and some other large, plant-eating animals, are *ruminants*. Ruminants have complicated stomachs, usually made up of four parts. When food is first swallowed, it goes to the *rumen*, where it is broken down before being returned to the mouth as *cud* for a thorough chewing, called *rumination*. When it is swallowed a second time, it goes to the *reticulum*. From there it passes to the remaining areas of the stomach until the digestive process is complete. Sheep, deer, antelopes and giraffes are all ruminants.

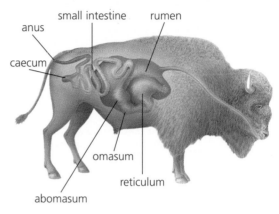

small intestine

rumen

anus

caecum

omasum

reticulum

abomasum

food and begins to digest it. When the food is well chewed, the animal swallows it.

The chewed food passes down the gullet into the stomach. Muscles in the gullet push it forward by squeezing together behind it and relaxing in front of it in a movement called *peristalsis*. The *stomach* is a strong muscular bag, where the food is mixed with acid and digestive juices called *enzymes* which kill any bacteria and help to break the food down.

Partly digested food passes from the stomach into the small intestine. Enzymes in the wall of the intestine, and from a gland called the *pancreas*, continue digestion. They break the food down into small molecules – sugars, fats and amino acids (from proteins). These can pass easily through the wall of the small intestine into the blood. First these molecules are carried to the *liver*. The liver stores excess foods and changes some poisons into harmless substances. From the liver the digested food is carried in the blood to all parts of the body, where it provides energy and materials for growth and repair.

The parts of food which cannot be digested, such as fibre from vegetables and fruit, pass into the large intestine, which is made up of the colon, rectum and anus. The undigested food remains in the colon for between 12 and 36 hours. Water and salt are removed from it during this time. It then passes into the rectum and out of the anus as faeces.

Dinosaurs

Dinosaur records
Longest dinosaur
Diplodocus 27 m
Tallest dinosaur
Brachiosaurus 12 m
Smallest dinosaur
Compsognathus
75–91 cm; 3 kg
Heaviest dinosaur
Ultrasaurus 120
tonnes

• At the time of the first dinosaurs, the dry land of the Earth formed one supercontinent, called Pangaea. Animals were able to reach all parts of Pangaea, and so, when it split to form separate continents, dinosaurs lived on all of them.

• The word 'dinosaur' means 'terrible lizard'. More than 1000 different kinds of dinosaur are known, and more are found every year.

Dinosaurs were prehistoric reptiles. Many of them were very big, larger than any other animal that has ever lived on dry land, but some were as small as chickens. Dinosaurs inhabited the Earth for about 160 million years, and the last ones became extinct 65 million years ago – more than 60 million years before the first human-like creatures appeared.

For most of the time that dinosaurs lived the climate worldwide was fairly warm and there were few high mountains. Most dinosaurs lived on broad plains by slow-flowing rivers.

The first dinosaurs appeared over 220 million years ago. They were quite small and ran on their hind legs. The main difference between them and their ancestors was that their legs were 'under' their bodies (like human legs), instead of sticking out to the sides. This gave them greater control over their movement and meant that they could reach food and escape from enemies more easily. They fed on insects and other small prey. They survived and thrived and many different kinds developed, some of which were very large.

Main dinosaur groups

Scientists have divided dinosaurs into two main groups based on the shape of their hip bones. One group is called the Saurischia, which means 'reptile-hipped', because reptiles such as crocodiles have their hip bones arranged in this way. The other group is called Ornithischia, which means 'bird-hipped', because the hip bones of birds are arranged like this. There are differences between the two groups. The reptile-hipped dinosaurs had their teeth in the front of their mouth. Though they could bite and nip and tear, they could not really chew their food. All the flesh-eaters were saurischians. The bird-hipped dinosaurs had their teeth at the back of their mouth, and often had a horny beak at the front. They

◄ Fossilized dinosaur footprints like this one are one of the clues that help scientists find out how big dinosaurs were and how fast they moved.

could grind up tough plant food. All the bird-hipped dinosaurs fed on plants (although none ate grass, which did not appear until after the dinosaurs had died out). Many of them had armoured plates of spikes or bone on their bodies.

Warm-blooded or cold-blooded?

Scientists do not know whether dinosaurs were warm-blooded or cold-blooded. A cold-blooded animal like a lizard needs heat from the Sun to keep warm and to become active. Warm-blooded animals, such as birds and human beings, can be active all the time. Because the Earth's climate as a whole was warmer, it would have been easier for cold-blooded dinosaurs to remain active for much of the time. The big dinosaurs could stay warm through the night because it would take a long time for their massive bodies

▶ Dinosaurs such as *Tyrannosaurus* (opposite) and the slimmer, more ostrich-like *Stenonychosaurus*, walked on their hind legs, rather like chickens.

▼ Other dinosaurs walked on all fours. Some of these, like *Triceratops* (top), had a bony neck shield. Others had horns or plates of bone on their bodies, like *Scolosaurus* (bottom).

to lose heat. This may be one reason many of the dinosaurs were so large. Smaller dinosaurs may have kept themselves warm by being active, or they may have been truly warm-blooded.

The end of the dinosaurs

No one knows why the dinosaurs disappeared. For 160 million years, new kinds of dinosaur had evolved as others became extinct (died out). But at the end of the Cretaceous period, 64 million years ago, the dinosaurs disappeared quite suddenly.

Scientists do not know exactly how long this disappearance took. It could have happened in a short time or it might have taken as much as a million years. Many other kinds of animals became extinct at about the same time. Most of them were large, although many smaller species disappeared as well, including the mosasaurs, the last of the great sea reptiles.

So what happened? One suggestion is that this was a time of great changes in climates and habitats. The dinosaurs could not adapt to the new conditions and just died out. Another theory, popular today, is that the Earth was hit by a giant meteorite. The impact sent up a huge cloud of dust which blacked out the Sun. This lowered the temperature, and plants and animals died off everywhere. Only those that could withstand the cold were able to survive and spread out across the Earth.

Discovering dinosaurs

Everything that we know about dinosaurs has been learned from fossils. If scientists discover a new kind of dinosaur, they study its bones to work out the shape of its body and how it walked and ran, and what other kinds of dinosaur it is related to. The position in which some dinosaur fossils have been found tells us that certain kinds of dinosaur lived in herds.

There may also be other fossils such as leaves, insects or shells that tell us about the climate and the plants that lived at the time. Dinosaur teeth often survive, and can tell us whether the animal ate flesh or plant food. A few dinosaur eggs have been found, and these suggest that some kinds of dinosaur, like most reptiles alive today, did not look after their young.

All these things help us to understand the dinosaurs, but there are many things that we do not yet know about them. What colour were they? How noisy were they? How long did they live? Scientists are still searching for the answers to these questions.

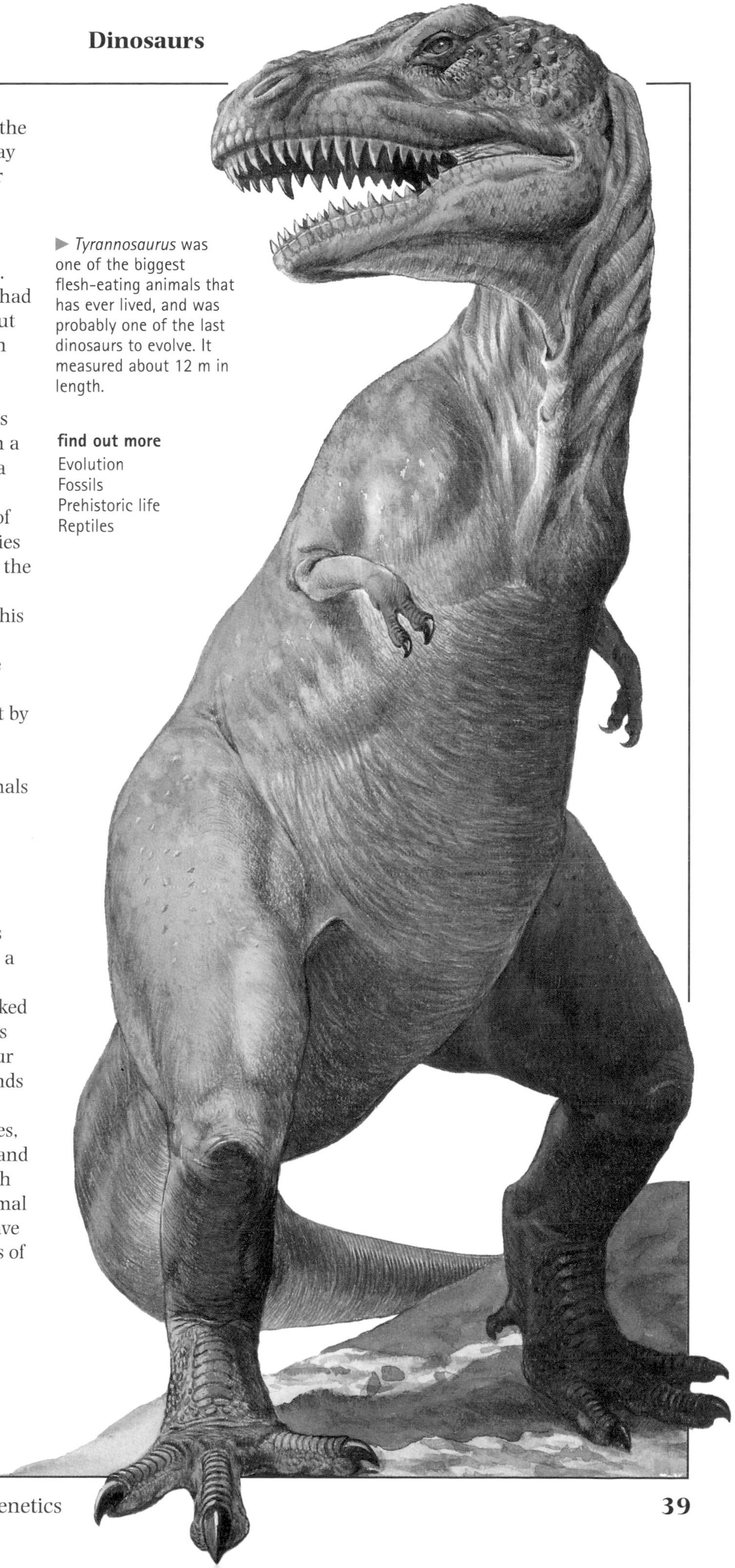

▶ *Tyrannosaurus* was one of the biggest flesh-eating animals that has ever lived, and was probably one of the last dinosaurs to evolve. It measured about 12 m in length.

find out more
Evolution
Fossils
Prehistoric life
Reptiles

Dogs

There are more than 100 different breeds of domestic dog that live and work with people all over the world. All these breeds are descended from just one ancestor, the wolf. But there are many other kinds of wild dog as well, including wolves, foxes, jackals and coyotes.

All dogs are hunters, and most can run fast. They possess very good senses of sight, hearing and smell. All these senses help them to hunt, but for most wild dogs sight is the most important. They use their excellent sense of smell to communicate with each other. Like all hunters, dogs are intelligent animals.

Wild dogs

Wolves live in family groups called packs. The pack centres around a breeding pair who mate for life and can contain up to 20 members. Wolf cubs are fed and cared for by all members of the pack. Because wolves hunt together, they can kill animals that are much larger than themselves. They feed mainly on deer, usually preying on the weaker ones.

After a kill, a wolf can eat about 9 kilograms of meat in one go.

Jackals live in the warm, dry climates of Africa, Asia and south-eastern Europe. They usually feed on rodents, lizards, birds and insects, although they can hunt larger animals, such as small antelope. They also scavenge, eating the leftovers from lions' kills and sometimes human rubbish. Jackals are hunted as a pest, and because of this and the destruction of their habitat, Simien jackals are now an endangered species.

Unlike wolves and jackals, most foxes live alone, although cubs remain with both parents for the first few months of life. Foxes live in most regions north of the tropics, and in Australia, where they were brought by European settlers. They eat a wide variety of foods, such as small rodents, insects, and even woodland fruits. Some foxes kill game birds, but most do not, and they are often useful in destroying pests.

Domestic dogs

Dogs were the first animals to be domesticated and they are now found wherever there are human beings. All dogs have the same biological make-up and behaviour patterns, and they are capable of inter-breeding, producing cross-breeds (mongrels). Dogs look so different from each other because they have been bred by people for different purposes.

The domestication of dogs began towards the end of the Palaeolithic period (Old Stone Age), about 10,000 BC. It may have come about because wolves were attracted to scavenge for food at the rubbish tips left by the humans of that time. Perhaps wolf cubs were sometimes found and kept, and

▲ Jackals usually live in pairs. Once mated, they will stay together for life. This family of golden jackals from Ngorongoro, Tanzania, has just caught a flamingo. Cubs stay with their parents for at least six months.

• There are 2 different kinds of wolf, 21 different kinds of fox and 4 different kinds of jackal. Other wild dogs include the African hunting dog, the bush dog, the dhole and the racoon dog.

◄ Red foxes survive well in many different conditions. They are found over more of the world than any other carnivore (flesh-eater). They have recently taken to living in towns and cities, although they are still wary of humans.

Dogs

Record-breaking dogs

Tallest
Irish wolfhound, shoulder height about 110 cm
Heaviest
St Bernard, weight about 100 kg
Smallest
Chihuahua, weight under 1 kg
Fastest
Over long distances, saluki; over short distances, greyhound

• The coyotes of North and Central America have, like foxes, adapted well to living beside their human neighbours and now inhabit regions which were formerly the territory of wolves.

• Dingos were probably introduced into Australia by the Aborigines 5000 to 8000 years ago. They can sometimes be tamed, and become affectionate pets.

find out more
Animal behaviour
Mammals

▼ Humans have bred dogs for all sorts of jobs, so the variety among breeds of dog is enormous.

people discovered, as they grew up, that they could be useful. It took several thousand years to transform wild wolves into tame dogs, but by about 5000 years ago dogs were playing an important part in people's lives.

Wolves defend their territories, so dogs could easily be trained to guard the human living places that they shared. Wolves also defend their pack. In ancient times large breeds of dog were used for fighting in battles, defending their human pack. They have also been trained to hunt, to herd animals and to fight each other for sport. In the days before central heating, people sometimes kept themselves warm by cuddling small dogs, called 'lap dogs'.

A useful companion

Today our understanding of dog behaviour has made it possible for us to train dogs to do new and more difficult jobs. Some are guide dogs for the blind and hearing dogs for the deaf. Others are trained as sniffer dogs to help track down terrorists and drug dealers, and others as disaster dogs, to find people buried in avalanches or the rubble of earthquakes. They are also used for experiments in laboratories. Large dogs, such as bouviers, were often used to pull small carts, and in the Arctic sledge dogs are vital to human survival.

▲ All breeds of domestic dog are descended from the wolf. Wolves were once widespread but have now been reduced to small populations in Europe, Asia and northern regions of North America.

Because dogs are hunters, they like chasing games and must have plenty of exercise. Pet dogs that are not given enough to do can sometimes become bored and destructive. Like wolves, dogs are pack animals, attached to and obedient to their group. One reason why they have taken so well to domestication and human companionship is that people have become their 'pack'.

chihuahua   Afghan hound   cocker spaniel   Airedale   cairn terrier   bulldog   chow chow   bouvier

Dormice *see* Mice, etc. • **Dragonflies** *see* Insects • **Dromedaries** *see* Camels

Ducks, geese and swans

Wildfowl is the name we often give to the large group of birds which includes ducks, geese and swans. They are usually found near water and have webbed feet, short legs and oily feathers.

There are three main groups of ducks. Dabbling ducks, such as mallards and shovelers, filter the surface water to find food. Diving ducks, such as pochards and sawbills, search below the surface for vegetable matter. Perching ducks, including muscovies and the colourful mandarins, nest and perch in trees.

True geese are found only in the northern hemisphere. They graze on grass and other vegetation. They spend much of the year in large flocks and many make long migrations. Grey geese include the greylag goose, which is the ancestor of many farmyard geese. Among the black geese is the Canada goose of North America, which has also been introduced into Britain.

Swans feed on grass and water plants. They have long, flexible necks which allow them to reach plants up to a metre below the surface of the water. Several kinds of swan are long-distance migrants. Bewick's swans fly from the Russian tundra to north-west Europe for the winter. The closely related whistling swan breeds in Arctic Canada and winters in the southern USA.

• Swans and geese keep the same mate all their lives, but ducks usually find a new mate each season. For this reason, male ducks usually produce richly coloured feathers each breeding season, in order to attract females. Female ducks, and geese and swans of both sexes, are more plainly coloured throughout the year.

• There are 147 different kinds of ducks, geese and swans.

find out more
Birds

▼ The mute swan (right) with a cygnet (young swan), the greylag goose (centre) and the shoveler duck (left). Geese are smaller than swans but bigger than ducks, and spend less time in water than either of them.

Ears

We use our ears to detect sound. Other mammals have ears similar to ours, and birds, reptiles, amphibians and fishes are able to hear sounds, although their ears do not work in quite the same way. Some insects have ears on their legs or body.

The part of the ear that you can see is called the outer ear. Its funnel shape enables it to collect sound waves and focus them on a thin membrane called the *eardrum*. The eardrum then vibrates, which causes the group of tiny bones (*ossicles*) that are attached to it to vibrate too. This vibration in turn moves the liquid in a snail-shaped chamber inside the inner ear called the *cochlea*. Inside the cochlea, sensitive hair cells, attached to nerves, send messages to the brain. These nerve messages enable us to hear and understand sounds, including speech and music.

The ear and brain working together can separate sounds depending on how loud or soft, and how high or low pitched they are. Having two ears helps us to tell where a sound is coming from, as it will be louder in one ear than the other unless it is straight ahead or immediately behind. The lowest sounds humans can hear are at 20 vibrations a second (a low hum) and the highest are 20,000 vibrations a second (a high-pitched hiss). Some animals, such as bats, can hear higher sounds than these.

Your sense of balance is controlled in the inner ear. Any change in the position of your head causes fluid in three *semicircular canals* to move. This information is then passed by nerve cells to the brain.

find out more
Bats
Senses

▼ Sounds make the eardrum vibrate. This makes the ossicles vibrate, which makes sensory hairs in the cochlea vibrate. These send nerve impulses to the brain where we hear the sound.

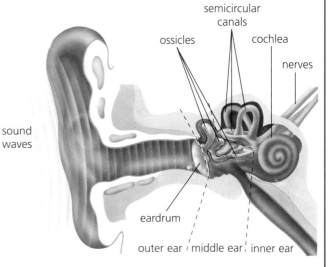

semicircular canals

ossicles

cochlea

nerves

sound waves

eardrum

outer ear / middle ear / inner ear

Ecology

Ecology is the science that studies how plants and animals live together, and how they affect, and are affected by, the world around them. Scientists who study ecology are called ecologists.

▲ The leafy treetops (the *canopy*) of the Amazon rainforest are the richest habitat in the world. Ecologists use many techniques to study the canopy at close quarters. This biologist is using climbing equipment to reach an artificial nest hole high above the rainforest floor. He is holding an egg that was laid in the hole.

There are many complicated links between plants, animals and their environment. For example, thrushes eat the berries of trees like the mountain ash. The seeds inside the berries pass through the gut and come out in the thrush's droppings. The seeds then grow into new trees, which produce more berries. In bad summers, the trees produce few berries and the thrushes cannot feed so many young, so the number of thrushes falls. The trees therefore depend on the thrushes, and the thrushes depend on the trees, and both are affected by the weather.

To take another example: voles and lemmings are small mammals that live in northern countries. In years when the weather is kind, they have lots of young. This means there is also plenty of food for the owls that hunt them. The owls are therefore able to raise lots of young, and so the next year there are many more owls. They kill large numbers of their prey (the voles or lemmings), so the numbers of the voles and lemmings drop. The numbers of the owls, voles and lemmings therefore rise and fall in a way that is closely linked together. The job of an ecologist is to try to understand how these kinds of natural processes work.

Careful study

Ecologists often need to make careful measurements as part of their study. So, in the example above, they might count the number of owls and lemmings, measure how much grass is available for the lemmings to eat, and record careful details of the weather. Putting all these measurements together, they can begin to understand the processes that control the numbers of the different animals.

Animals are also linked together in what ecologists call food chains. In the sea, for example, tiny plants float near the surface, using the Sun's energy to make their own food. These are eaten by tiny, floating animals. These tiny animals are eaten by small fishes, which are eaten by bigger fishes. Then the bigger fishes might be eaten by a penguin, and the penguin might be killed and eaten by a seal. The animals involved therefore depend on other animals lower down the food chain, and all their food comes originally from the energy of the Sun captured by the tiny plants.

Living spaces

Ecology also shows how plants and animals depend on all the features that make up their surroundings, called their *environment*. So cacti and rattlesnakes, for example, are adapted (suited) to live in the desert environment, and this is described as their *habitat*. To survive there, they must have ways of saving water, keeping

• Deserts, rainforests and coral reefs are examples of ecosystems. Even a small pond can be an ecosystem in itself.

thrush

seeds in droppings

new tree

mountain ash

▲ The thrush and the mountain ash depend on each other. The thrush needs the tree's berries for food, while the mountain ash needs the thrush to spread its seeds.

• The complex way in which the lives of plants and animals are linked together is called the 'balance of nature'. A change in one part of an ecosystem often causes changes elsewhere that bring the system back into balance. However, the ecosystem is very finely balanced. If the changes are too great, it may break down altogether.

lion

grass and leaves

grazing animals

▲ A lot of energy is lost with each link in a food chain. This can be pictured as a 'food pyramid'. On the Serengeti plains of East Africa, a single lion (0.2 tonne) needs a population of 350 grazing animals (55 tonnes) from which to feed if the number of grazing animals is going to stay stable. The grazing animals, in turn, need 6500 tonnes of grass and leaves from which to feed.

cool, coping with the shifting sands and avoiding their enemies. The spines of the cacti, for example, stop them from being eaten by animals, while rattlesnakes have a special way of moving over the hot ground. The lives of all the plants and animals in any habitat depend on each other and on their surroundings. The living system that ties them all together is called an *ecosystem*.

Lessons for life

One of the lessons of ecology for humans is that, if we disturb the balance of nature in any ecosystem, the result is difficult to predict but is often damaging to us and to many other kinds of living things. The story of what happened to the Californian sea otter illustrates this.

Early in the 20th century, so many sea otters were hunted for their fur that they almost died out. Sea otters feed on smaller animals, including sea urchins. With the sea otters gone, huge numbers of sea urchins built up in California, USA. They soon used up all the large seaweeds, called giant kelps, on which they fed. Underwater 'forests' of kelp were replaced by bare, rocky areas called 'sea urchin barrens'. But the kelp forests were the breeding place for many fishes off the Californian coast. When the kelp disappeared, the fishes died out. The fishing industry had to close down, and so did the canning factories. Hundreds of people lost their jobs – and all because the killing of sea otters had disturbed the balance of nature.

▶ The disappearance of the Californian sea otter, as a result of hunting, had unexpected effects that led to many fishermen and factory workers losing their jobs.

Eels

Eels are long, thin fishes that look a bit like snakes. Freshwater eels are among the commonest fishes in the rivers of Europe and North America – but they have one of the most remarkable life stories in nature.

Adult eels live for nearly 20 years in rivers and streams, until they are fully grown. Then they start on a long journey down to the sea and out across the Atlantic Ocean. They swim for thousands of kilometres until they reach a warm, calm area of the ocean called the Sargasso Sea. Here they lay their eggs. Then they die.

When the tiny eel larvae (young) hatch from their eggs, they do not look like eels. Instead they are transparent, with ribbon-shaped bodies. The larvae drift in the ocean, and eventually currents carry them to the coasts of Europe and North America. Here they change form and turn into small eels, called *elvers*. The elvers travel up rivers and streams in great numbers, sometimes wriggling overland through wet grass. When they reach their feeding grounds, they gradually grow into adult eels.

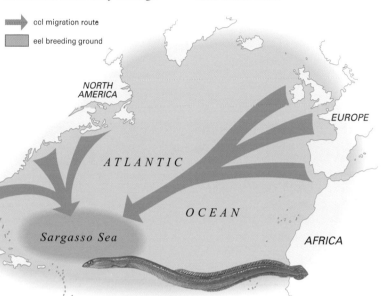

→ eel migration route

▨ eel breeding ground

NORTH AMERICA

EUROPE

ATLANTIC

OCEAN

Sargasso Sea

AFRICA

• There are many kinds of eel as well as the common eel. Most spend all their lives in the oceans. They include the large conger eel, the fearsome moray eel, and the deep-sea gulper eel, which has a huge head and jaws, and a long, thin body.

◀ Young eels hatch from their eggs in the Sargasso Sea, then drift to the rivers of Europe or North America. Many years later the adult eels return to the Sargasso Sea to lay their eggs and die.

find out more
Fishes
Migration

Eggs

An egg is the cell made by a female animal which, when joined with a sperm from a male, can grow into a new animal. Almost all animals produce eggs.

Most animals lay their eggs. The eggs are normally covered with a protective layer of jelly or a tough shell. Fish eggs are laid in the sea or in lakes and rivers; turtle and crocodile eggs are laid in sand; frogs' eggs (frog-spawn) are laid in ponds; and birds' eggs are laid in nests.

The eggs of many insects have shells that are very nutritious. The young insects eat these shells as soon as they hatch. Insects also often lay their eggs on a plant that will be the food of the young when they hatch.

Birds' eggs

A bird's egg has several layers. On the outside is a hard shell for protection. Inside are membranes (thin walls) and the *albumen* (white). This is a watery substance that cushions and protects the growing chick. Right in the centre of the egg is the *yolk*, which is made of fat and proteins to feed the growing chick.

When an egg is laid, if it has not been fertilized, a chick will not develop. If it has been

◀ Many kinds of cuckoo lay their eggs in other birds' nests and leave the 'foster mother' to rear the young birds. This cuckoo egg has been laid in a dunnock's nest. When the cuckoo hatches, it will push the other eggs out of the nest so that it can eat all the food brought by the foster mother.

fertilized, a chick grows. It must then be *incubated* (kept warm), and the parents do this by taking turns to sit on the eggs. At the end of the incubation period the young bird hatches by breaking out of the egg.

find out more
Birds
Cells
Growth and
  development
Mammals
Sex and reproduction

▼ A cross-section of a chicken's egg. The chick sits on a ball of food called the yolk. This floats in a jelly called the albumen which gives it moisture. The shell gives protection and also lets oxygen pass through to the growing chick.

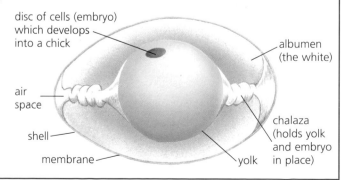

disc of cells (embryo) which develops into a chick

albumen (the white)

air space

shell

membrane

yolk

chalaza (holds yolk and embryo in place)

Some differences between Asiatic and African elephants

Asiatic elephant
Length up to 6.4 m
Weight up to 5000 kg
Small triangular ears
Rounded back
Two bulges on forehead
Smaller tusks (no tusks in female)

African elephant
Length up to 7.5 m
Weight up to 7500 kg
Large, rounded ears
Concave (hollowed) back
Rounded forehead
Larger tusks

• In the past there were many species of elephant living over most of the world apart from Australia. Some were much bigger than the elephants of today. They have now all died out.

find out more
Conservation
Mammals
Prehistoric life

Elephants

Elephants are the biggest living land animals. There are two kinds of elephant: one lives in Africa, south of the Sahara Desert, the other in southern and South-east Asia. African elephants are rarely tamed, but Asiatic ones are often used to move heavy objects and for ceremonial processions.

Both kinds of elephant live in herds. Each herd is led by an old female, who is followed by her young offspring and her adult daughters and their families. The elephants in a herd remain together for many years.

Young males leave the herd when they become sexually mature, at the age of about 12 years. Males may form herds, but these may change from one day to the next.

▼ A young African elephant uses its trunk to pull up grass as it grazes.

Trunks, tusks and teeth

An elephant's trunk is formed from its upper lip and nose. It is boneless but muscular, and has one or two finger-like points at the tip. The elephant uses it to smell, breathe and drink. The trunk is also used as a hand. It is strong enough to break down a branch and delicate enough to pick a single fruit the size of a raspberry.

An elephant's tusks are really just big teeth. Most of a tusk is made of dentine (ivory). Elephants use their tusks for defence and in feeding. They continue to grow throughout the elephant's whole life, so one with very large tusks is likely to be an old animal.

To support its large bulk, an elephant needs huge amounts of food. An adult elephant eats about 150 kilograms of grass, leaves, twigs and fruit each day. Such tough food needs to be thoroughly chewed, and elephants have grinding teeth in the back of their mouths. When these teeth get worn out, they are replaced by another set which pushes in from behind. When the last of these wears out, the animal becomes weakened by lack of food and dies of starvation or disease.

▲ Elaborately decorated tame Asiatic elephants take part in a festival at Trichipooram in southern India.

Elephants in danger

Present-day elephants are endangered because humans are increasingly taking over the land where they live and also killing them for their valuable ivory tusks. Ivory has always been much prized for carving and making luxury items. Poaching for ivory has severely reduced the populations of African elephants, and many African countries now ban its export.

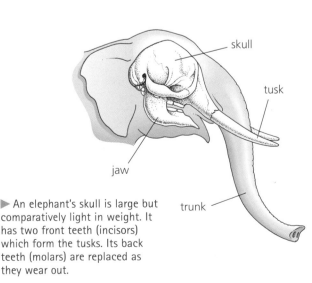

skull
tusk
jaw
trunk

▶ An elephant's skull is large but comparatively light in weight. It has two front teeth (incisors) which form the tusks. Its back teeth (molars) are replaced as they wear out.

Evolution

Evolution is the way new kinds of plants, animals and other living things come into being as a result of many small changes over a long period of time.

Evidence from fossils in ancient rocks shows that over millions of years new kinds of creature have appeared. At the same time some of the old forms, such as the dinosaurs, have become extinct (died out).

In the 19th century the biologists Charles Darwin and Alfred Russel Wallace proposed a theory to explain how evolution takes place. The theory of *natural selection* (sometimes called the 'survival of the fittest') argues that the offspring of an animal or plant that are most likely to survive are those best suited to the present conditions and best able to compete for the things they need, such as food, water, light and space. These 'fit' individuals are also more likely to produce offspring, which in turn will inherit the special characteristics that made their parents fit. In a few generations that kind, or species, of plant or animal will contain more fit creatures and will gradually change. This may eventually lead to the development of a new species.

► Darwin's theory of evolution can explain the development of the pterosaur, an ancient flying reptile that lived over 150 million years ago. It began as a four-footed lizard (1). Over millions of years, small folds of skin developed between its feet, which enabled it to glide from tree to tree (2). Over many more generations the folds, and the bones and muscles supporting them, grew to form wings (3).

Charles Darwin

Charles Darwin (1809–1882) was an English scientist who came up with the revolutionary idea that plants and animals evolve (change) over time by a process called natural selection.

Darwin did not pursue his interest in plants and animals seriously until he became friends with the cleric and botanist John Stevens Henslow at Cambridge University. On Henslow's recommendation, he was invited to join the naval survey ship, HMS *Beagle*, as the ship's naturalist, and as a gentleman companion for the captain. He set sail on 27 December 1831.

For Darwin, the most important part of the five-year voyage was visiting the Galapagos Islands, west of Ecuador. He noticed that the finches on the different islands were closely related, yet the finches from any one island were different from all the others. Also, they all looked similar to a type of finch that lived on the South American mainland. Darwin decided that some of the finches must have first reached the islands accidentally and then evolved separately on each island.

Darwin concluded that, over many generations, living things evolve and adapt to their environment through a process of natural selection. This means that only the best-adapted survive and produce offspring. In 1859 Darwin published his theories in his book *The Origin of Species*. It was very unpopular because it went against the belief that all kinds of plants and animals were created by God and had not changed since. Today few people doubt the logic of Darwin's arguments.

► Darwin became ill a year or so after his return from the *Beagle* voyage, and his health gradually declined. For the last 40 or so years of his life he was an almost permanent invalid, but he continued his scientific work until his death.

Eyes

Eyes are like living cameras. They focus light from surrounding objects to form a picture which your brain can understand. Your eyes give you a clear, moving, three-dimensional, coloured picture of the outside world. They allow you to recognize things by their colour, shape and brightness.

Human eyes work in much the same way as the eyes of other vertebrates (animals with backbones). At the front of the eye is a clear, round window called the *cornea*. Tears from tear glands keep it moist and clean and, together with your eyelids and eyelashes, help protect it from damage.

Behind the cornea, the lens focuses light onto a layer of light-sensitive cells at the back of the eye called the *retina*. The images focused on the retina are detected by these cells, which send messages along nerves to the brain. The brain

interprets them and converts them into what you see. On the retina directly opposite the lens is a patch called the yellow spot. This is the part which is most sensitive to colour.

Just behind the lens is a sheet of muscle called the *iris*. This is the coloured part of the eye. It has a round hole at its centre called the *pupil*. In bright light the iris muscles contract and make the pupil smaller, to stop too much light entering the eye. In dim light the muscles relax and the pupil opens and lets in more light.

▲ The eye of a typical mammal, cut away to show its internal structure. The image at the back of the eye is upside-down. The brain interprets it the right way up.

We need two eyes, because with only one eye we would find it difficult to judge distance and depth.

find out more
Brains
Nervous systems
Senses

Colour vision

Not all animals can see colours. Those that can detect colour include humans, apes, monkeys, some birds and reptiles, and amphibians and fishes. Insects have good colour vision. In fact some insects, including bees and ants, can see ultraviolet light that is invisible to us.

In vertebrates with colour vision, special cells called *cone* cells at the back of the eye are sensitive to colour. Different types of cone cell are sensitive to different colours. These colours are red, green and blue, and all the colours you see are mixtures of these three primary colours.

▼ This tarsier from the Philippines has large eyes, which help it to hunt for food in the night. In the dark its pupils expand to collect as much light as possible. In daytime or bright light its pupils become very small.

▼ The eyes of insects, like this dragonfly, are made up of hundreds of narrow, light-sensitive tubes. Each tube cannot make a proper picture but all the tubes together can. Eyes like this are called *compound eyes*. Crustaceans, such as crabs, also have compound eyes.

Fishes

Fishes are cold-blooded animals with backbones, gills and fins. They all live in water, and can be found in ponds, lakes and rivers, and in the sea. Most sea-dwelling fishes live in the shallows close to land and in the surface waters of the great oceans.

Fish records
Largest
Whale shark, up to 18 m long, over 40 tonnes
Largest freshwater fish
Arapaima, up to 3 m long and 200 kg
Smallest fish
Pygmy goby, up to 11 mm long, 4—5 mg
Fastest
Sailfish, about 109 km/h
Largest number of eggs
An ocean sunfish contained an estimated 300 million eggs

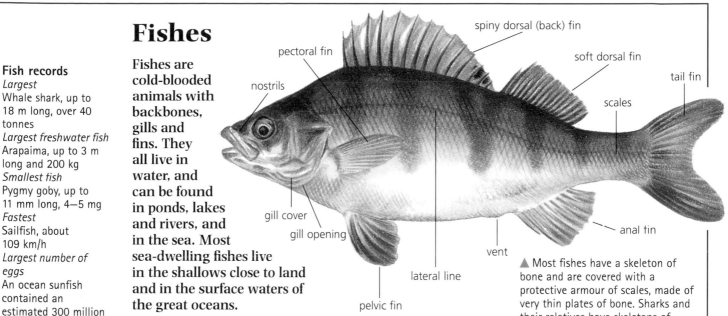

spiny dorsal (back) fin
pectoral fin
soft dorsal fin
nostrils
tail fin
scales
gill cover
gill opening
vent
anal fin
lateral line
pelvic fin

▲ Most fishes have a skeleton of bone and are covered with a protective armour of scales, made of very thin plates of bone. Sharks and their relatives have skeletons of gristly cartilage. There are about 21,000 different kinds of fish, of which more than 20,000 are bony fishes.

▼ A lionfish hunts among a school of cardinal fish off the coast of Borneo in Indonesia.

Water is about 800 times as dense as air, so fishes have to work hard to swim through it. The fastest fishes, such as mackeral and tuna, have long, tapering, streamlined bodies and crescent-shaped tails. Slower-moving fishes, such as tench and carp, have rounder bodies and squarer tails. Some kinds of fish, such as rays and turbot, live on the sea-bed. They have flattened, camouflaged bodies and are slow-moving, and so they are hard to see.

When they swim, most fishes use powerful zigzag muscles which move their bodies and tails from side to side. Only very few kinds paddle along with their fins, which are mainly used for balancing, stabilizing and braking.

Fishes do not sink in the water even when they are totally still. Most kinds of fish have an internal *swim bladder*, a gas-filled bag which holds them up. When they swim, all their energy goes into pushing forwards through the water. Other animals have to use some of their energy to stay up when swimming.

The life of a fish

Most fishes are born from eggs, although a few kinds are born live. Usually fishes have a definite breeding season. Some, like eels and salmon, migrate long distances to reach the best place to spawn (produce eggs).

In most fishes the females lay huge numbers of eggs. These are almost always fertilized by the male outside the female's body. It is unusual for fishes to take any care of their young, although some produce smaller families and do look after them. Among these it is often the male, rather than the female, that cares for the young.

Fishes that live in fresh water or near the seashore usually lay eggs that sink to the bottom, where they may be hidden among plants. Fishes that live in the open sea generally lay very tiny eggs that float. When these hatch, they become part of the sea plankton. Very few of these tiny fishes survive, as they provide food for many other animals.

The senses of a fish

Fishes have the same basic senses as we do: sight, hearing, touch, smell and taste.

Most fishes have eyes on the sides of their heads, which gives them all-round vision. Many fishes have good colour vision. This is important in the

Fishes

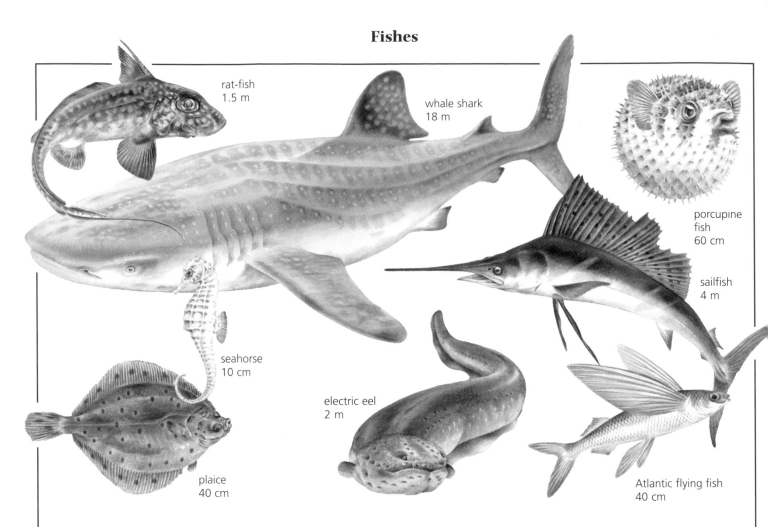

rat-fish
1.5 m

whale shark
18 m

porcupine
fish
60 cm

sailfish
4 m

seahorse
10 cm

electric eel
2 m

plaice
40 cm

Atlantic flying fish
40 cm

courtship of some species, and in others that can change their colour to match their surroundings.

Fishes' ears are inside their skulls. They work as organs of balance as well as of hearing. Sounds travel very well under water, and many fishes make noises to communicate with each other.

Many fishes live where there is little light, so their sense of touch is important. A fish's main organ of touch is its *lateral lines*. These are a series of very sensitive nerve endings that lie just below the skin along the fish's sides. Any movement in the water is detected by the lateral lines, and the fish can tell that there is an enemy or a possible meal nearby. Some fishes have 'whiskers', called *barbels*, round their mouth which they use to explore the sea- or river-bed.

Most fishes have a very good sense of smell and can recognize the scents of other fishes. Many use smell to find their prey. Some, if they are injured, release a special substance from their skin into the water. When other fishes of the same species smell this, they swim quickly to safety. Salmon probably use their sense of smell when they migrate back to the stream where they were born to lay their eggs.

The sense of taste is related to the sense of smell. Some fishes have large numbers of taste buds in and around the mouth.

Fish food

Freshwater fishes eat all sorts of food, including water plants, snails, worms, insects and their larvae, and other fishes. Some fishes, like grey mullet, suck up mud from the river-bed and

digest the tiny creatures that live there. They are called detritus feeders.

In the open sea, where there are no rooted plants, most fishes are carnivores (flesh-eaters), feeding on a wide range of animals including shrimps, squid and other fishes. Others – including some very large fishes, such as basking sharks – are filter-feeders. The fish takes a mouthful of water, closes its mouth and pushes the water out over its gills. The plankton is filtered out by special combs called gill rakers, which are attached to the gills, and the fish then swallows it.

Some of the strangest of all fishes are those that live deep in the ocean. Most have huge mouths and teeth and some can swallow creatures bigger than themselves, as they have few chances to feed and may not eat again for months.

▲ Almost half the 43,000 species of animals with backbones (vertebrates) are fishes. Some of the species that have evolved are very strange indeed.

Ferrets *see* Weasels and their relatives

Fleas and lice

Fleas and lice are wingless insects that live as parasites on other animals. A parasite is an animal that lives by feeding on another animal (its host), usually causing it harm in the process.

Fleas live on mammals and birds. They have backward-pointing hairs and sharp claws, so they are difficult to dislodge if the host tries to scratch them away. They can walk slowly, but if they need to move quickly they jump using their powerful back legs.

Fleas feed on blood: the bite itches because the host's body reacts against chemicals in the flea's saliva. Fleas lay their eggs in the nests or beds of their hosts. The young that hatch out look like tiny worms. Some fleas breed at the same time of year as their hosts, so that when the host's young are born there is a crop of fleas ready to live on them.

Two kinds of insect are called lice. One group is the bird lice, or biting lice. As their name suggests, they are found mainly on birds, where they nibble at feathers and skin. The other group is called the sucking lice. These feed on mammals' blood. They cling closely to their hosts with gripping claws. Unlike fleas, lice attach their eggs, called nits, to the host's hair.

The human louse may live in hair, or in clothes which are close to the skin. Head lice can easily be spread where people are living close together, and perhaps sharing combs. It is important to control lice, as they can carry diseases.

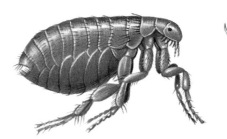

▲ A flea (left) and a louse (right) magnified many times. They often carry diseases which they can pass on to their host when they bite it to suck its blood.

- Fleas have been known to jump more than 30 cm, about 200 times their body length.

- There are about 1750 different kinds of flea, about 2600 kinds of biting louse and about 250 kinds of sucking louse.

find out more
Insects
Pests and parasites

Flies and mosquitoes

Flies are among the insects we see most often. Many people think they are dirty and spread disease, but this is not true in most cases.

Flies can be very useful. Some live on decaying plants and animals and can change these dead things into chemicals which help the soil. Other flies are pollinators, helping plants to spread by carrying pollen from flower to flower.

Flies vary a lot in shape. One big group, which includes mosquitoes and midges, have slender, soft bodies. A second group, which includes horseflies, have short, hard bodies. The third and largest group includes houseflies and hoverflies. All of these have short, well-armoured bodies.

Flies spend the first part of their lives as larvae, which are called grubs or maggots. Some larvae live in the flesh of dead mammals or birds; some hunt tiny soil-dwelling animals; and other kinds live in water. A few types of larva burrow into the skin of cattle and other animals.

Mosquitoes are small flies, many of which carry diseases. A person can catch the disease if they are bitten by the female mosquito (the male feeds only on plants). Malaria is one of the most serious diseases spread by mosquitoes: it probably kills a million people, mostly children, each year – more than any other infectious disease. Taking anti-malarial tablets reduces the chances of infection if you happen to be bitten.

▲ A male mosquito's feathery antennae are sensitive to sound vibrations: they are its 'ears'. This mosquito is about six times its actual size.

▲ Crane-flies are usually known as daddy-long-legs.

- There are at least 90,000 different kinds of flies and mosquitoes.

- All flies have two pairs of wings. The front pair are used for flying, while the tiny hind pair are probably used to help with balance.

find out more
Insects
Pests and parasites

◀ The male horsefly sucks nectar from plants but the female sucks blood from horses, cattle and humans. This horsefly is about three times its actual size.

Flowering plants

Flowering plants are the most successful group in the plant kingdom. There are nearly 250,000 species, found in every part of the world. The smallest is the least duckweed, a tiny water plant scarcely 1 millimetre across. The largest flowering plants are trees, which can grow to 95 metres or more in height.

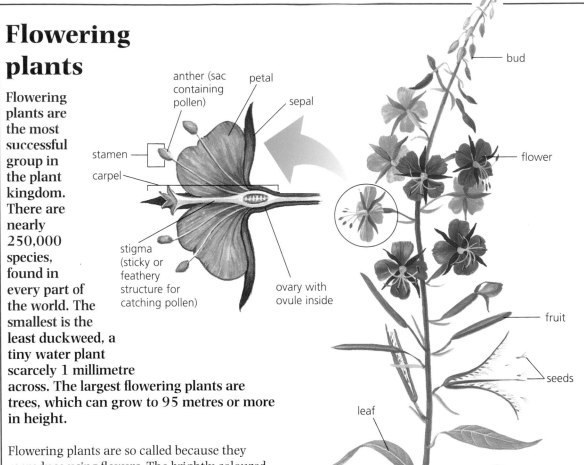

anther (sac containing pollen)

petal

sepal

stamen

carpel

stigma (sticky or feathery structure for catching pollen)

ovary with ovule inside

Flowering plants are so called because they reproduce using flowers. The brightly coloured and scented flowers that we see in the countryside and in gardens are only one type of flower. Flowers of plants like grasses are small and inconspicuous, while other flowers, like the catkins of some trees, do not look like typical flowers at all.

Parts of a flower

There are many different types of flower, but they all have the same basic structure. Each part of the flower is designed to do a particular job.

The outermost parts of a flower are the leaf-like *sepals*. They protect the flower when it is in bud. *Petals* are the most visible part of the flower. They are often coloured and scented to attract insects.

Stamens contain the plant's male sex cells. The male sex cells are tiny *pollen grains*, which are spread from plant to plant. *Carpels* contain a flower's female sex cells, the *ovules*. These are like the plant's eggs. An ovule cannot grow into a new plant until it has linked up with a pollen grain.

In most flowering plants, each flower has both stamens and carpels. However some plants, such as oak and willow trees, produce separate male and female flowers.

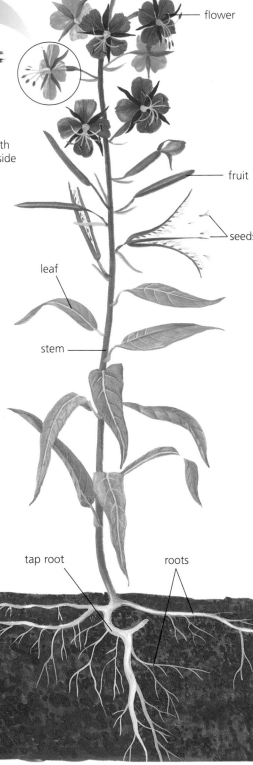

bud

flower

fruit

seeds

leaf

stem

tap root

roots

◄ Parts of a flowering plant.

Leaves make food for the plant, using energy from sunlight, carbon dioxide from the air, and water. This process is called photosynthesis.

Stems are the main support of the plant. They also contain the tubes that carry water up from the soil to the leaves, and food from the leaves to other parts of the plant.

Roots anchor a plant and draw in water from the soil, along with the minerals needed for growth. The main (tap) roots may reach far down into the soil, while a mass of fibrous roots spreads out through the soil surface.

● All flowering plants belong to a group of plants known as *angiosperms*. There are two subgroups. One includes grasses, lilies and orchids. The other includes most common flowers and all broadleaved trees.

Orchids

Orchids are the largest flowering plant family, with at least 18,000 known species. Most grow in tropical countries, partly because insects to pollinate them are more common there. Some tropical orchids grow high above the ground, with dangling roots which absorb water and minerals from the rain.

Pollination and fertilization

A flower's pollen has to combine with an ovule on the same or another flower before the ovule can become a seed. But first, the pollen has to get to the ovule. This process is called *pollination*.

Most flowers have developed ways to ensure that pollen from one flower is carried to another flower of the same species. There are two main ways that this transfer takes place: by animals and by wind.

▲ Hazel trees have both male and female flowers. These catkins are the male flowers. They are shedding a dust of pollen into the air, to be carried away on the wind.

▲ A honeybee on a swamp rose (the smaller insect is a hoverfly). The bee is searching for nectar, which it uses to make honey. Bees also collect pollen, in special 'pollen baskets' on their back legs. The yellow dust on this bee's legs is pollen.

Insects are the most common animal pollinators, especially bees and butterflies. Insects are attracted to a flower by its colour and smell. They come in search of *nectar*, a sweet, sugary liquid found in a tiny cup at the base of each petal. While the insect is searching for nectar, pollen grains stick to its body. Some of this pollen is then left behind on the next flower the insect visits.

Wind-pollinated flowers need to be in a position to catch the breeze. Their pollen grains are small and light so that they can float on the slightest breeze. Pollen is produced in huge quantities, because each grain only has a very small chance of reaching another flower. Spreading, feathery structures (*stigmas*) on each flower act like nets to catch pollen drifting past in the breeze.

Once a pollen grain has reached another flower of the same kind, it grows a tiny tube down into the carpel, to join up with an ovule inside. Once it reaches the ovule, the two cells combine. This is *fertilization*. Once the ovule has been fertilized, it can start developing into a seed. At the same time, the part of the carpel around the seed (the *ovary*) swells to form a fruit, such as a fleshy berry, a nut or a dry capsule.

Life cycles of flowering plants

Each growing season, a flowering plant increases in size until it is ready to flower and set seed. *Annuals* are plants which flower once and die within a year, leaving their seeds to survive the winter or dry season. *Biennials* also flower only once, but take two years to do so. In the first year they produce stems and leaves, and store food, usually in their roots. The shoot then dies but the root survives underground. In the second year, the plant uses the stored food to grow a new shoot. It then flowers, releases its seeds, and dies. *Perennial* plants flower year after year, each year storing food in their roots for the next year's growth.

▼ The talipot palm has one of the largest flower clusters in the world. It only flowers once, then dies, exhausted by the effort of producing so many flowers.

Food chains and webs

All living things need food to stay alive. Green plants make their own food, using sunlight, water and carbon dioxide from the air. Animals cannot do this. They have to get their food either by eating plants or by eating other animals. So the plants and animals which live in an area are all connected to each other by the way that they get their food.

When green plants use sunlight to make their food, they are storing some of the Sun's energy. Plant-eating animals (*herbivores*) get this energy directly from the plants, by eating them. The herbivores are eaten in turn by flesh-eating animals (*carnivores*), which may themselves be eaten by other carnivores. But even for these animals, the energy in their food comes originally from plants.

The simplest food chains

The plants and animals which live in an area are all linked together by energy and food, forming a food chain. In a garden, caterpillars may eat a cabbage plant. Some of the energy stored in the cabbage is passed to the caterpillars. If a thrush then eats a caterpillar, some of the energy is passed to the thrush. Cabbage, caterpillar and thrush are all links in the food chain.

There may be only two links in a food chain, as when a horse eats grass. If large carnivores feed

▲ In a few food chains, a plant is the carnivore rather than an animal. The Venus fly-trap grows in soils that are short of certain minerals that the plant needs for growth. It gets the extra minerals by catching and digesting insects in its specially adapted 'fly-trap' leaves.

on smaller carnivores, there may be three or more links in the food chain. If we use our original example, a fox may kill and eat the thrush. Then the food chain has four links.

Food webs

In real life, food chains are rarely quite so simple. Many different food chains may interconnect to form a food web. All living things are linked together in food webs. A change in the numbers of animals or plants in one part of a food web (for example, as a result of pollution) can affect all the other parts of the web. Human activities often disrupt natural food webs, putting the whole living system (the *ecosystem*) out of balance.

• Scavengers and decomposers are an important part of most food chains. *Scavengers* are animals such as vultures, hyenas and some kinds of beetle. They feed on dead plants or animals. *Decomposers* are certain types of bacteria and fungi, which break down the remains of dead animals. This puts minerals back into the soil and helps new plants to grow.

◄ A food web. In many meadows, field voles eat grass. Kestrels often hunt and kill the voles. The grass, voles and kestrels make a simple food chain (red arrows). Other kinds of animals also feed on grass – cattle, sheep, deer, rabbits, snails and grasshoppers. In turn, there are many carnivores which feed on these herbivores.

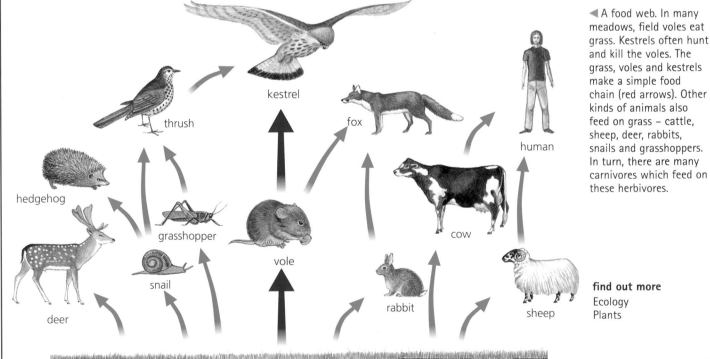

kestrel

thrush

fox

human

hedgehog

grasshopper

cow

vole

deer

snail

rabbit

sheep

find out more
Ecology
Plants

Forests

A forest is a large area of land covered mainly with trees and undergrowth. Some forests, such as the great Amazon rainforest, have existed for thousands of years. Vast areas of forest have been destroyed by human activity, but others have been planted, and a fifth of the world's land is still covered by forest.

Forests create their own special environment. At ground level they are generally shady and cool, because the crowns of the trees shade the forest floor. The air is still, because the trees shield the forest interior from the full force of the wind. The days are cooler and the nights warmer, making the forest a sheltered place for wildlife.

Types of forest

Forests can grow wherever the temperature rises above 10 °C in summer and the annual rainfall exceeds 200 millimetres. Different climates and soils support different kinds of forest.

Rainforests thrive in the humid tropics. The weather is the same all year round – hot and very rainy. The rainforest has such a dense tangle of vegetation that it is often difficult to distinguish the various layers. Plants, including some orchids, even grow perched on the trunks and branches of trees and fallen logs. There are many different kinds of tree: a small patch of forest may contain over 100 different tree species.

Deciduous forests are found in temperate climates, where it is cool in the winter and warm in the summer. There are many fewer different types of tree than in the rainforest. The main trees are deciduous, which means that they shed their leaves in winter (or in the dry season in some regions). The trees also produce flowers, fruits and nuts at particular times of the year. When the trees are leafless in springtime, the forest floor receives plenty of light, so smaller plants may grow and flower.

Coniferous forest is found further north and higher up mountain slopes than any other kind of forest. The main trees are conifers (cone-bearing trees) such as pines. Most are evergreen (they keep their leaves all year), with needle-like leaves coated in wax to reduce water loss. Conifers can survive drought and the freezing of soil water in winter. Their branches slope downwards, so that snow easily slides off. Some conifer forests have arisen naturally, but others have been planted by people, in order to grow wood for timber, paper and other uses.

The importance of forests

Forests have important functions. Like all green plants, trees absorb carbon dioxide from the air, and release oxygen. So forests are major suppliers of the oxygen humans and all animals need, and they help stop the levels of carbon dioxide in the air from rising too high. This is important because high levels of carbon dioxide in the air cause global warming. ♦

▼ The structure of a forest
In the rainforest shown here, the larger trees form an almost continuous *canopy* over the roof of the forest. A few very tall trees, called *emergents*, grow through the canopy into the sunlight above. Below the canopy smaller trees and young saplings form the *understorey*. Below them are shrubs and briars, and on the *forest floor* a layer of smaller shrubs and plants. Each layer of the forest has its own animal communities.

1 harpy eagle
2 macaw
3 spider monkey
4 cock of the rock
5 tree boa
6 sloth
7 ocelot
8 poison frog
9 capybara
10 coral snake

▲ Autumn in a deciduous forest in Utah, USA. As the leaves die, the chemical that makes leaves green is broken down, and they turn many different colours. By shedding their leaves for the winter or during the dry season, the trees are able to save energy and greatly reduce the amount of water they need.

A forest acts like a giant sponge, absorbing rainfall and only gradually releasing it into rivers. By holding the water back, the forest prevents disastrous flooding further downstream. Forest cover also prevents the soil being eroded (worn away) and silting up rivers and lakes.

Forests provide shelter and food for many animals. Leaves, flowers, fruits, seeds and nuts are food for insects, birds and small mammals, which in turn are food for larger birds and mammals. The moist forest soil has its own community of worms, centipedes, beetles, ants, and the eggs and larvae of many insects.

Forests provide people with an enormous range of important products. For large areas of the world, firewood is the main source of energy. Wood is used for house frames, shipbuilding, papermaking, packaging, fencing and many other purposes. The fruits and nuts of forest trees provide food and spices. Forest trees also provide oils for cooking and industry, syrups, resins, varnishes, dyes, corks, rubber, kapok, insecticides and medicines such as antibiotics.

Vanishing forests

Every year the world consumes 3 billion cubic metres of wood. In Britain and much of north-west Europe, human beings have cut down most of the trees, although today in these regions people are planting more trees than they are cutting down. In tropical regions, an area of rainforest the size of a football pitch is cut down every second.

Many rainforests grow on very poor soils. Most of the goodness of the land is locked up in the plants. When rainforest is cut down and burned, the remaining soil is often too poor to support crops for long. The soil is easily washed away in the tropical rains, and silts up lakes and reservoirs. The area may even turn to desert.

Forests, especially tropical rainforests, contain a huge variety of plants and animals. If forests are cut down at the present rate, around half the world's remaining rainforests will have vanished by the year 2020, with the loss of around one-tenth of all the species on Earth. Scientists and others are campaigning to stop this destruction of the Earth's resources.

find out more
Conservation
Flowering plants
Plants
Trees and shrubs

▼ **Rot and decay**
When trees die or parts of the tree fall, they quickly rot. Woodpeckers drill holes in soft rotting trunks, and bark beetles excavate tunnels under the bark, allowing fungi to enter. Fungi break down the tree's tissues and absorb the nutrients. When the fungi die, they in turn are broken down by bacteria. The dead plant material gradually crumbles into the soil, and the nutrients it contained escape into the soil, to be taken up by new plants.

wireworm (beetle larva)

snail
slug
fungi
centipede
woodlouse

cockchafer larva

millipede

bacteria

pot worms

mite
and
springtail

earthworm

ants

Fossils

find out more
Dinosaurs
Evolution
Prehistoric life

• The word 'fossil' comes from a Latin word meaning 'dug up'.

▼ How fossils are formed.

Fossils are the remains of animals and plants that died long ago. By studying fossils, we can discover what kinds of animals and plants lived on the Earth many millions of years ago.

Most fossils show us only the hard parts of the ancient animal or plant. These include the materials making up the shells and bones, which do not rot away. The commonest fossils are ancient sea shells. This is because most of the rocks that contain fossils were formed in the sea, and shellfish have always been common. Other sea creatures, such as fishes and corals, can also be found quite often.

The rarest and most exciting fossils show us the soft parts of long-dead animals. These include insects trapped in amber, a resin that oozes from certain trees. Very occasionally, whole mammoths, deep-frozen for thousands of years in the soil of Siberia, are discovered. Fossilized animal droppings, footprints and burrows are also sometimes found.

1 When a dead animal sinks to the sea-bed, scavenging animals and bacteria soon remove its flesh.

2 Sediments pile on top of the remains, and gradually minerals replace the chemicals in the bones.

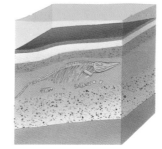

3 Water is squeezed out and the rock becomes hard and compact.

4 Millions of years later the rocks become dry land. Rain, wind or the sea wear them away until the fossil is exposed.

Frogs and toads

• There are about 2600 different kinds of frogs and toads. They live in moist habitats throughout the tropical and temperate areas.

• You will find a feature box on the life cycle of a frog under Amphibians.

find out more
Amphibians
Camouflage
Newts and salamanders

Frogs and toads are amphibians, like newts and salamanders, and like them, they spend the early part of their lives in water. The main differences between frogs and toads and other amphibians is that they do not have tails and that they have powerful hind legs for swimming and jumping on land.

Frogs and toads are all good swimmers and on land most move about by leaping in a zigzag pattern. Many are active mainly at night. All adult frogs and toads feed on other small animals. Some catch their prey with long, sticky tongues, others leap and grab insects in their mouths. Some have sucker pads on their toes which make them very good climbers. There is no very clear distinction between animals called frogs and animals called toads. True toads usually have drier, wartier skins than frogs.

Frogs and toads have small lungs, but they also breathe partly through their moist, soft skins. If their skin were to dry out, they would suffocate.

In the breeding season males call to attract their mates. Each kind of frog or toad has its own

▶ This fire-bellied toad shows its bright underside to an enemy to signal 'red for danger'. The poison in its skin tastes so unpleasant that other animals soon learn not to touch it.

croaking or trilling song and many males have throat pouches which they can blow out to increase the volume of sound.

Frogs and toads are preyed on by many other animals for food. Some protect themselves by camouflage – their colouring matches their background so they cannot be seen easily. They also have powerful poison glands in their skin, and some of these are brightly coloured to warn enemies to steer clear. The poison-dart frogs of South America have such powerful poison in their skin that the native people used to tip their arrows with it.

Fruit

Scientists use the word fruit to describe the part of a flowering plant where the seeds develop. The fruit encloses and protects the seeds as they develop, then helps to spread them when they are ripe. Different kinds of fruit spread their seeds in different ways.

Fruits we eat, like apples, oranges and bananas, are called 'soft fruits'. Other fruits form dry or papery containers round the seeds. These are 'dry fruits'.

Animals spread the seeds of some fruits. They eat the sweet flesh of soft fruits, and drop the seeds inside, or else the seeds pass through the animal's gut without being digested and fall to the ground in the droppings. Fruits that are sticky or have tiny hooks ('burs') hitch rides by sticking to an animal's fur or feathers.

Many plants use the wind to spread their fruits and seeds. Sycamore, maple and ash all have winged fruits, while dandelions and thistles have parachutes of hairs. Poppies and orchids produce tiny seeds, light enough to float on the breeze when they are shaken from the fruit.

flower

fruit

▲ Stages in the growth of a tomato from a fertilized flower. After fertilization, the flower petals wither and drop off, and the ovary (the part containing the seeds) swells into a round green fruit. When the seeds are ready, the tomato ripens from green to red and develops a sweet flavour.

Plants growing near a river or the sea, like the coconut palm, may use water currents to carry their fruits. Other fruits, such as pea pods, split or burst open to release ripe seeds.

• Nuts are not soft fruits, but they contain rich food and many animals eat them or bury them to eat later.

• 'Vegetables' like tomatoes, cucumbers and peppers are really fruits, and contain seeds inside.

find out more
Flowering plants
Plants

Fungi

Fungi are an important group of living things. They get their food by breaking down the tissues of living or dead plants or animals. Many fungi can only be seen with a microscope. A few, such as mushrooms and toadstools, grow much larger.

Most fungi grow as networks of tiny threads called *hyphae*. These spread through the material on which the fungus is growing, and absorb food. Fungi reproduce by way of microscopic *spores*, which act like tiny seeds. They are produced in a part of the fungus called the *fruiting body*.

Mushrooms and toadstools

Mushrooms and toadstools are the fruiting bodies of some kinds of fungus. They are usually umbrella-shaped, with a thick stalk and a broad cap. Beneath the cap there are hanging curtains called gills or pores, where the spores develop. Many kinds of mushroom are edible, but only two are grown widely for food. There is no easy way to tell a poisonous mushroom from an edible one, so it is safest to eat only ones from the shops.

Fungi in the soil

One vital role of fungi is to decompose (rot) dead plants, animals and their wastes. Fungi break the dead matter down into chemicals that are recycled back into the soil. Other fungi are essential partners in the growth of trees and other plants. They live in or around their roots in a mass called a *mycorrhiza*. The roots protect the fungi, which in turn provide food for the plants.

◄ Four of the many kinds of fungus.
1 Field mushroom 2 Shaggy ink cap
3 Leaf rust 4 Bracket fungus.

Helpful fungi
Yeasts are a type of fungus used in making bread, beer and wine. Other fungi produce penicillin and other antibiotics, which attack bacteria harmful to humans.

• There are about 70,000 known species of fungus. Although they are plant-like, fungi are usually put in their own kingdom, the Mycota.

• Some fungi grow so fast that a single spore can produce more than 1 km of hyphae in 24 hours.

find out more
Algae
Plants

Genetics

The young of all living things are similar to their parents. A mother rabbit always has baby rabbits; apple seeds always grow into apple trees. Children inherit characteristics from their parents. Genetics is the study of how these characteristics pass from one generation to the next.

▼ Identical twins (below) look very alike because they have the same genetic material. They both grow from a single egg that splits after it has been fertilized. Non-identical twins look less alike: they grow from different eggs and do not have the same genetic material.

Different kinds of animal or plant are obviously different from one another. But there are also smaller differences between animals or plants of the same kind. Pansy flowers, for example, can be many different colours, while female spiders are bigger than males. Some of these differences are due to *heredity*. Heredity is the inheritance of characteristics from our parents.

Environmental influence

Not all of an individual's characteristics are inherited. Some things about us are purely the result of our environment and upbringing. Reading and writing, for example, are things that we learn. We also have characteristics that result from a combination of both inheritance and environment. Our body shape is partly due to heredity, but how much we eat can have a major influence on it.

Chromosomes and genes

Characteristics are passed on from parents to their offspring through structures called *chromosomes*. These are found inside the cells that make up the body. Each cell carries inside it a complete blueprint of the whole plant or animal.

There are a number of chromosomes in a cell, and they are arranged in pairs, one from each parent. Different animals and plants have different numbers of chromosomes – humans, for example, have 46 chromosomes arranged in 23 pairs.

Arranged along the length of each chromosome, like beads on a necklace, are units called *genes*. Each gene carries information about a particular characteristic of the plant or animal. A human gene, for example, might carry information about eye or hair colour. More complicated features such as body shape are the result of many genes working together.

Every chromosome in a cell is different. But the chromosomes of a pair carry genes for the same characteristics. So for each characteristic we have two genes.

Genes produce the different characteristics of an animal or plant by controlling the making of proteins. There are thousands of different proteins, each with an important job in the body. Each gene controls the making of one or a small number of these proteins. Which genes your cells have influences which proteins are produced.

Passing on genes

A new animal or plant is made by the joining together of two special cells, called *sex cells*, one

Sex chromosomes

The difference between males and females is in their chromosomes. One particular pair of chromosomes is responsible for deciding the sex of a plant or animal. In humans, this sex chromosome can be either X or Y. A person with two X chromosomes will be female; someone with an XY chromosome pair is male.

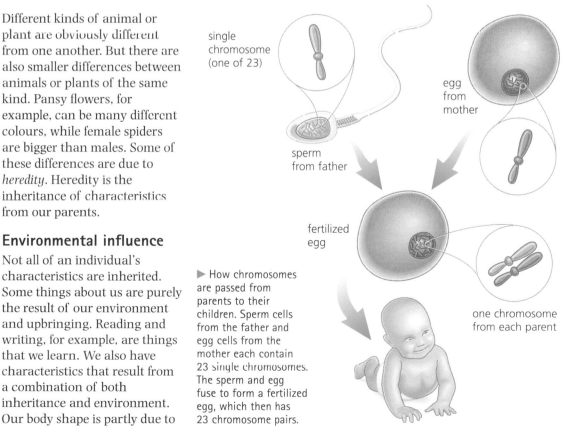

single chromosome (one of 23)

sperm from father

egg from mother

fertilized egg

one chromosome from each parent

▶ How chromosomes are passed from parents to their children. Sperm cells from the father and egg cells from the mother each contain 23 single chromosomes. The sperm and egg fuse to form a fertilized egg, which then has 23 chromosome pairs.

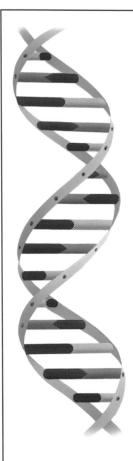

◀ Genes are made from a chemical called *DNA*. Each gene is one section of an enormously long DNA molecule. DNA is like a long ladder, twisted into a spiral. The 'rungs' of the ladder carry coded information that body cells can use to make proteins.

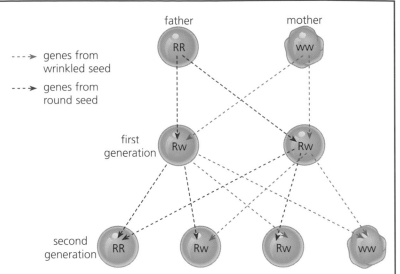

---> genes from wrinkled seed

---> genes from round seed

▲ Breeding wrinkled-seed pea plants with round-seed pea plants. The offspring from these plants produce only round seeds. But if two of the offspring plants are bred together, they produce a mixture of round and wrinkled seeds, with about three times as many round seeds as wrinkled seeds. The symbols inside the peas show the gene pairs in the plants in each generation. The round-seed gene, R, is dominant over the wrinkled-seed gene, w.

from each parent. Sex cells are different from all other cells in the body, in that they do not have a complete set of chromosomes. Each sex cell contains only half the normal number of chromosomes, one from each chromosome pair. When two sex cells join, the resulting cell (which becomes the new animal or plant) has the normal number of chromosomes – half from the mother and half from the father.

Dominant genes

The characteristics we inherit are passed on through pairs of genes. But if we have two genes for each characteristic, which one does the body use? Or does it use both?

Of the two genes we inherit for any characteristic, one of them is usually *dominant* over the other. For example, in pea plants, there is a gene that controls whether the pea seeds are round or wrinkled. A plant with a round-seed gene from each parent will have round seeds, and a plant with a wrinkled-seed gene from each parent will produce wrinkled seeds. But the round-seed gene is dominant over the wrinkled seed gene. So if a plant inherits a round-seed gene from one parent and a wrinkled-seed

gene from the other, it will produce only round seeds.

This dominance of one gene in a pair over another means that animals and plants carry within their chromosomes many 'hidden' genes. These do not show as characteristics because they are masked by a dominant gene.

Genetic engineering

Scientists' improved understanding of genetics has made it possible to 'engineer' living cells to give them useful properties. For example, microbes (tiny, single-celled creatures such as bacteria and yeasts) can be modified (changed) to produce medically useful vaccines, hormones and other chemicals that are normally made only in the human body.

Genetic engineering has many important uses, but many people are worried by it. They do not like the idea that one day babies may be 'designed' to have specially chosen characteristics.

find out more
Biotechnology
Cells
Proteins
Sex and reproduction

▶ Dolly the sheep, born in 1997, was the first large animal to be cloned from an adult. A cell from a Dorset sheep (Dolly's mother) was injected into an unfertilized sheep's egg that had had its nucleus removed. The two cells were fused using a spark of electricity. The new cell was placed in the womb of a third sheep, where it grew into Dolly. Dolly has exactly the same genetic material as her mother.

Giraffes

The giraffe is the tallest animal in the world. It has very long legs and a very long neck. The top of its head can be more than 5.5 metres above the ground. This astonishing height enables it to feed on leaves high up in trees.

Giraffes live in the wooded grasslands of eastern and southern Africa. They have pale-coloured coats covered with large reddish-brown blotches, in a variety of patterns. On the top of their heads there are one or two pairs of short, hair-covered horns. Despite the length of their necks, they only have seven neck bones – the same as most other mammals. Giraffes also have very good eyesight which,

with their height, enables them to watch out for danger over great distances.

Giraffes live in small groups of females and their young, led by a single male. Males fight to become the leader of a group by swinging their heads at each other. Female giraffes usually give birth to a single offspring.

The only animals that hunt giraffes are humans and lions. Giraffes defend themselves by kicking with their hooves, and can run at speeds of nearly 50 kilometres per hour.

• The okapi is the only close relative of the giraffe. This shy creature lives in the rainforests of Congo, in central Africa. It was not discovered by Europeans until the beginning of the 20th century.

◀ The necks of giraffes are so long that they have to spread their legs in order to reach down to a waterhole to drink.

find out more
Digestive systems
Grasslands
Mammals

Glands

Glands are special organs which make chemicals that the body needs, such as sweat and saliva. Some produce chemicals called hormones that control important life processes.

The chemicals produced by glands are called *secretions*. Sweat glands in your skin produce sweat, and salivary glands in your mouth produce saliva. Glands in the stomach and the wall of the gut make digestive enzymes which break down food.

Hormones

Hormones are made and stored in glands, and released into the blood to be carried around the body. Humans have many different hormones.

If hormones do not carry out their job properly, it can lead to disease. In the disease called diabetes, the pancreas does not produce enough insulin, so the blood is too sugary. People who are diabetic feel very tired unless their condition is treated.

Hormones also control important processes in other animals. For instance, in insects and crabs hormones control moulting (shedding of the external skeleton). In plants hormones control growth and flowering.

▶ Some of the main glands of the human body. Other mammals have similar glands.

When boys reach puberty, their **testes** start to produce a hormone called testosterone, which causes their voice to get deeper, their beard to grow and their sex organs to enlarge.

The **pituitary gland** produces hormones which control growth, reproduction and milk production, and hormones that control other glands.

The **thyroid gland** produces thyroxine, which increases the activity of certain body processes.

The **pancreas** produces the hormone insulin, which makes the liver absorb sugar from the blood.

Women's **ovaries** produce sex hormones that control the menstrual period and egg production.

find out more
Digestive systems
Growth and
 development
Sex and reproduction

Grasses

Grasses are the most successful flowering plants, with about 900 different kinds. Extensive grasslands cover more than a fifth of the world's land surface.

◀ Flowering cocksfoot grass, and a diagram of one floret. Grass flowers use the wind to spread their pollen (male sex cells). The flowers consist of two or three florets. Each floret contains three anthers on long stalks, and two feathery stigmas. The anthers contain pollen, which they release into the wind. The stigmas catch pollen from other flowers. (In a real floret, stigmas and anthers do not appear at the same time.)

The grass stem is divided into sections by nodes (joints), with a leaf attached at each node. Part of the leaf wraps round the stem to form a sheath: the rest forms a long, thin blade. Most grasses are *herbaceous*. This means that they do not have woody stems and die right down to the ground after flowering. An exception is bamboo, which has woody stems up to 40 metres high. Many grasses are annual: they live for a season, produce a large quantity of seeds, and then die.

Grasses for food

In the prairies of North America and the savannah of Africa, grasses are food for huge herds of grazing animals. For people too, grasses are a very important food. Wheat, maize, rice, sugar cane and bamboo are all types of grass. About half the energy that humans around the world take in from food comes from wheat, maize and rice alone.

Why grasses are successful

Grasses survive because they grow quickly and soon produce seed. Their tall, thin shape means that they fit easily between other plants. If flattened, the stem will grow to bring the flower stalk upright again. Many grasses have short shoots, which lie close to the ground, so they can go on growing even if heavily grazed.

find out more
Cereals
Flowering plants
Grasslands
Tundra

Grasshoppers and crickets

Grasshoppers and crickets are insects that have existed since before the dinosaurs. Most are plant-eaters, though some feed on other small creatures. Almost all have huge hind legs, and can leap long distances.

• There are about 13,000 different kinds of grasshoppers and crickets.

• In the Old Testament of the Bible a plague of locusts is one of the plagues with which God punished the Egyptians to make them free the Israelites.

The animals that we usually call grasshoppers have short, rather thick antennae, so they are often known as the short-horned grasshoppers. Most are active during the daytime. Crickets have long, hair-like antennae which are much longer than their bodies. Many crickets are active at night.

Both grasshoppers and crickets 'sing'. Grasshoppers do this by rubbing a row of little pegs on their hind legs against a vein on the side of their forewings. Crickets rub their wings together. Only the males sing, courting females with a song that also warns off other males. Though it is easy to hear grasshoppers and crickets, they are usually so well camouflaged that they are very difficult to see.

Several kinds of large grasshopper are known as locusts. Most of the time they are harmless, but sometimes huge swarms of them gather together and destroy all the plants in an area as they feed. In the Middle East and Africa, locusts have been the most feared of all pests since farming began.

▶ A desert locust in flight. A large locust swarm may include 50 billion insects. They eat more in a day than all the people that live in London and New York.

find out more
Insects
Pests and parasites

Grasslands

Grasslands cover more than a fifth of the Earth's land surface. Grass is an important food for many animals, and humans grow and eat grasses like wheat and rice. Grasslands are of many different types, from the desolate steppe lands of Central Asia to the rich farmlands of North America.

• Some parts of the steppes of Russia, Kazakhstan and Mongolia are still roamed by camels and horses. Some of the animals are wild, others are used by nomads, who move about with their animals and live in large round tents made of felt (*yurts* or *gers*).

Grasslands have different names in different parts of the world. There are hot savannahs in East Africa, and veld in southern Africa. In North America there are the fertile prairies, in South America the dry pampas and chacos. The huge areas of grassland in Central Europe and Asia are called steppes, while in western Europe there are meadows and downlands.

Animals

All sorts of animals live on the world's grasslands. Australia's grasslands are home to many of the country's best-known animals, such as kangaroos, emus, kookaburras, and flocks of brightly coloured budgerigars, parrots and cockatoos. Some of the world's largest animals live on the savannahs of East Africa. They include giraffes, elephants and black rhinos.

The North American prairies are home to prairie dogs, small rodents that feed on the grasses. They eat the grass roots as well as the leaves, but they also collect and bury grass seeds, so helping to 'plant' new grasses. The rattlesnake is common in the grasslands of North America. The 'rattle' is in the tip of its tail: the snake rattles when it is disturbed or when coiled ready to strike.

find out more
Cereals
Grasses
Moors and heaths

▲ This rolling farmland in Washington State, USA, was once grassland. Much of what used to be prairie is now used to grow cereals such as wheat and maize (corn), which are themselves types of grass.

Most grasslands have a host of tiny animals, which are not usually seen. Earthworms, ants and beetles live in the soil, roots and leaves. Termites build large 'skyscraper' homes of soil in the grasslands, up to 7 metres high.

People and grasslands

Humans have changed grasslands in many parts of the world. Three hundred years ago, there were probably 60 million bison ('buffaloes') roaming the prairies of North America. Settlers from Europe shot bison for their meat and hides, and by the beginning of the 20th century fewer than a thousand were left. During the 20th century the prairies have been changed even more dramatically. The grasslands have been ploughed up and turned into rich farmland. Grasslands have been ploughed for crops in other regions too, such as the steppes of the Ukraine and Russia. In other areas, such as the pampas of Argentina, grasslands are used to graze beef cattle.

◄ Herds of giraffes, wildebeest, zebra and eland graze together on the East African savannah. Grazing in mixed herds enables the animals to help warn each other of danger from predators. Each animal eats different plants or parts of plants. Giraffes, for example, can reach leaves high on the acacia trees, and their tough mouths are not damaged by the trees' huge thorns.

Growth and development

▼ A sunflower plant begins as a seed. Roots grow (1), then a shoot pushes its way through the surface (2). The first leaves are special ones called cotyledons. Then the first full-sized leaves appear (3). At the top of the stem, a flower bud forms becoming a sunflower (4).

All living things are able to grow. Food is essential for growth, but the kind of food needed is different for plants and animals. Plants need only carbon dioxide gas from the air, water and mineral salts from the soil, and sunlight for growth. Animals need water too. But they also need many complicated foodstuffs like starches, proteins, fats and vitamins. They can only get these by feeding on the bodies of other animals or plants.

Although the foods used are different, the ways in which animals and plants grow are similar. To get bigger, living things make more cells. They change the food they take in into the material for new cells. These are formed when the cells that are already present divide into two. The faster this cell division goes on, the faster a plant or animal will grow.

In plants the cell division is concentrated at special growing points – often at the tips of twigs or roots. In a bud, tiny new leaves and sections of stem are produced out of new cells. They are then expanded to full size when the cells 'blow up', like minute balloons, with extra water.

Development before birth

In most animals and plants, a new individual is created when a sperm cell fertilizes (fuses with) an egg cell. Shortly after fertilization, the fused cell divides in two, and then these cells in turn divide, and so on. At this early stage, the growing fertilized egg is known as an *embryo*.

As the embryo grows, specialized cells begin to appear, and these develop into particular organs or types of tissue, such as muscle fibre. In mammals, once the organs have developed, the embryo becomes known as a *fetus*. In humans, this stage is reached after 8 weeks.

Soon after this the fetus is completely formed, and the rest of its time in the womb is spent in growing rather than changing form. In humans, birth occurs after about 40 weeks in the womb. In mice this period is only 20 days, while in large whales it lasts over a year.

Development after birth

Most mammals continue to grow fairly steadily after birth until they reach adulthood. During the course of this growth there is a period, known as *puberty* in humans, when the young develop into sexually mature adults, capable of reproducing themselves.

The life histories of some other animals, however, are marked by much greater changes. For example, a tadpole looks very little like the adult frog that it will grow into. In many insects, the young insect hatches out of its egg as a *larva* (grub), then changes into a *pupa*, before finally emerging as the adult form.

▼ Growth in the womb. Most mammals grow and develop in their mother's womb in the same way as the human baby shown here. Marsupial young, however, are tiny and have hardly started to develop when they are born, and young monotremes (the platypus and spiny anteaters) hatch from eggs.

The **umbilical cord** carries blood containing food and oxygen to the baby from the mother, and carries away waste.

6 weeks
0.5 cm

9 weeks
1.7 cm

The **amniotic sac** is a bag of fluid in which the baby floats inside the womb.

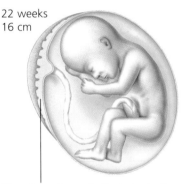

22 weeks
16 cm

The **placenta** roots the baby to the wall of the womb, and passes blood to the umbilical cord.

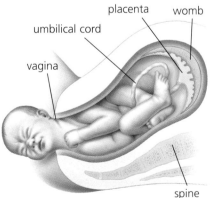

placenta
womb
umbilical cord
vagina
spine

▲ During birth (labour), muscles around the womb push the baby out, usually head first, through the vagina. Labour may take several hours.

Hair

Humans and other mammals are the only animals that have hair on their skin. The hair may be thin and patchy, as it is on humans, but most mammals have lots of it, tightly packed together.

A thick coat of hair, such as you find on a dog, is usually called fur. True fur is made of two sorts of hair: an inner layer of very fine hair that traps a layer of still air close to the animal's body, and an outer layer of thicker guard hairs.

Hair helps to control body temperature. It usually forms a waterproof, insulating coat that keeps down heat loss. Hair also gives an animal its characteristic markings. These are made out of hairs of different colours and lengths. An animal's markings are often used for camouflage and can be important in attracting a mate.

Animal hairs have other functions, too. For example, whiskers are specialized hairs that act as sensory organs for certain animals that are nocturnal (active at night). Porcupines and hedgehogs have sharp hairs called quills which are used for defence.

The structure of hair

Hairs grow out of a little pit in the skin called a *hair follicle*. The hair itself is dead and is made out of dead skin cells filled with a tough protein called keratin. Hairs do not last for ever. They grow for a while and then fall out and are replaced by new ones. The average life of a long hair on your head is between three and five years.

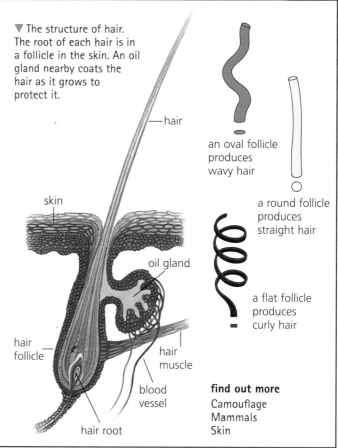

▼ The structure of hair. The root of each hair is in a follicle in the skin. An oil gland nearby coats the hair as it grows to protect it.

hair

skin

oil gland

hair follicle

hair muscle

blood vessel

hair root

an oval follicle produces wavy hair

a round follicle produces straight hair

a flat follicle produces curly hair

find out more
Camouflage
Mammals
Skin

Hearts

find out more
Blood

A heart is a muscular organ that pumps blood around the body. All vertebrates (animals with backbones) have hearts, as do many larger invertebrates, including earthworms, crabs, insects, snails and squids.

► How the heart works in mammals and birds.

pulmonary arteries carry blood from heart to lungs

head

right atrium

vena cava (vein) carries blood from body to heart

lung

pulmonary veins carry blood from lungs to heart

valves

right ventricle

left ventricle

valves

left atrium

aorta (artery) carries blood from heart to body

lung

body

In humans, the heart is in the middle of the chest. As with other mammals and birds, it is made up of four muscular chambers: two atria and two ventricles. Their strong contractions as they pump the blood are our heartbeats. The *pulse* is a wave of blood being pushed by a heartbeat along a blood vessel (tube) running from the heart.

These blood vessels are close to the surface at the wrist and neck, which are the best places to feel your pulse. Measuring your pulse rate tells you how fast your heart is beating.

How blood circulates

The left side of the heart takes in oxygen-carrying blood from the lungs. It then pushes it out along thick-walled blood vessels called *arteries* into very thin blood vessels called *capillaries*. The capillaries carry the blood to and from all parts of the body before joining up with thin-walled blood vessels called veins. The veins carry the blood, which now contains the waste gas carbon dioxide, back to the right side of the heart and the circulation begins again. The heart pushes the blood with the waste carbon dioxide out to the lungs and the gas is breathed out when you exhale.

Hamsters *see* Mice, etc. • **Hares** *see* Rabbits, etc. • **Hawks** *see* Hunting birds

Hedgehogs, moles and shrews

- There are 14 different kinds of hedgehog, 26 different kinds of mole, and 289 different kinds of shrew.

- A mole's fur is short and velvety, so that it can move either forwards or backwards in its tunnels and not rub its fur up the wrong way.

- Hedgehogs have about 3000 spines. They are actually thick hairs, and each one is hollow so all together they weigh very little.

find out more
Mammals

Although they look very different, hedgehogs, moles and shrews are in fact closely related.

Hedgehogs are covered with hundreds of spines. If a hedgehog feels threatened, it can roll up into a prickly ball and most enemies will leave it alone. Hedgehogs rest during the daytime, but in the evening they set out to look for food. They use their sense of smell to find insects, snails, slugs, grubs and frogs. They hibernate (sleep) through the winter.

Moles are active, burrowing creatures that spend most of their lives in long tunnels underground. They feed mainly on insects, worms and grubs in the soil. Moles are built for digging, with stumpy, powerful bodies and short legs. They have very poor eyesight but an acute sense of smell. They are also very sensitive to movement when they are underground.

Shrews are among the most active of all mammals, hunting small prey day and night. Their tiny bodies burn energy so quickly that some can starve to death if they go more than 4 hours without a meal. Shrews have stink glands on their sides which ensure that hunters such as weasels and cats leave them alone. Birds, however, have a poor sense of smell, and owls and birds of prey often hunt for shrews to eat.

◀ A hedgehog (left), a shrew (centre) and a mole. They all belong to a group of mammals known as insectivores. As the name suggests, insects form an important part of their diet. Hedgehogs grow up to 30 cm long, moles are 7–18 cm long, and shrews about 7 cm.

Hibernation

- Some animals that live in hot, dry places go into a kind of summer hibernation, called *aestivation*, if their food supplies disappear. Insects, snails, mammals and even some fishes aestivate.

- Many animals without backbones (invertebrates) hibernate. Snails and ladybirds cluster together in a safe place. Many butterflies come into buildings.

find out more
Animal behaviour
Animals
Migration

Hibernation is a kind of deep sleep in which some animals spend the winter. Most hibernating animals are small creatures that feed on insects, or other food that becomes scarce when the weather is cold.

Many cold-blooded animals, such as amphibians and reptiles, hibernate. These animals get their warmth from their surroundings. When the weather gets cold, their body temperature falls, they breathe slowly, and their heartbeat slows down. They become so sluggish that they can scarcely move. Because they are using hardly any energy, they can remain in this state for many months. They hibernate in safe places that are unlikely to freeze.

Warm-blooded hibernators

Almost all warm-blooded hibernators are mammals, including bats, hedgehogs and dormice. When they hibernate, they effectively cease to be warm-blooded: their temperature drops, and their breathing and heartbeat slow down.

Hibernating mammals feed a great deal in the summer, so that by the autumn they are carrying a lot of extra fat. This fat enables them to survive without feeding for months on end. Some hibernators, such as hamsters, make stores and wake up every now and then for a snack.

Many other mammals sleep for much of the winter. Badgers may not come out of their dens for days or even weeks, and bears snooze away the cold months. These animals do not really hibernate, because their bodies are functioning at a normal rate.

▼ This Daubenton's bat is hibernating for the winter in a cave. The outer parts of its body are so cool that water vapour in the air of the cave has condensed to form dew on its fur.

Hippopotamuses

There are two kinds of hippopotamus, or hippo, and both are found in tropical Africa. The common hippo is one of the largest land mammals after the elephant, but the pygmy hippo is much smaller.

Hippos feed on plants, and live in or near rivers and lakes. They are well suited to life in water. They have nostrils on top of their snouts that can close, which stops water getting up their noses, and spread-out toes to help them swim.

Common hippos have bulky bodies and huge heads and mouths, with large teeth like small tusks. They usually live in groups of 15 or so, but sometimes many more herd together. Common hippos spend most of the day resting in water.

Often they only leave their ears, eyes, and nostrils above the surface. They are good swimmers, and can walk along the bottom of rivers and lakes. At night they come on to the land, where they feed on grass and other plants.

Hippos have grey, almost hairless skins. The pinkish appearance of many hippos comes from a special oil that they produce to protect their skins. In the past, this led people to think that hippos 'sweat blood'.

▼ Common hippos can weigh up to 4.5 tonnes, and be over 3 m long. Pygmy hippos are less than 2 m long, and spend more time on dry land than common hippos. They live in rainforests, and are now very rare

find out more
Mammals

• Hippos are related to rhinos and horses. The word 'hippopotamus' means 'river horse'.

Horses

Horse records
Largest
One Percheron stood 21 hands (over 2 m), and measured over 5 m long
Smallest
The Falabella: adults about 70 cm (7 hands)
Fastest
69.2 km/h averaged over 400 m

• There are seven different kinds of horse, of which two (donkeys and horses) have been domesticated.

• A pony is a small horse, standing less than 142 cm (14 hands) at the withers (the high-point between its shoulder-blades).

find out more
Grasslands
Mammals
Zebras

For thousands of years, and in almost all parts of the world, people have used horses for riding, to pull heavy loads and for sport and leisure. Herds of wild horses still survive in some parts of the world – though all but the zebras of Africa are very rare.

Wild horses live in herds in open grassland, in Africa, the Middle East and Asia. They have sharp senses of sight, hearing and smell, and can detect hunting animals from a distance. They are strong runners, and use their speed over long distances to escape from danger. Herds of feral horses (domestic horses gone wild) survive in many places, especially in the Americas and Australia.

Domestication of horses

Horses were first domesticated by prehistoric people in central Asia over 6000 years ago. At first, they were used to pull lightweight chariots, but when larger breeds were developed, people began to ride. At first they rode without saddles or stirrups. The Roman cavalry were the first to use saddles regularly, and stirrups were only introduced into Europe from Asia in the 9th century AD. With the arrival of the horse collar at about the same time, people began to use horses to pull heavy weights.

During the Middle Ages, the armour worn by knights was so heavy that only very big horses could carry their weight. More lightweight, speedier horses were bred in the countries surrounding the Mediterranean Sea. These were the ancestors of the fast, slender-limbed horses such as the Arab breeds.

▲ The small, stocky Przewalski's horse comes from the steppes (grasslands) of Mongolia and western China. Herds are found in zoos and reserves, but it is probably extinct in the wild.

Human beings

find out more

Mammals
Primates

Today there are over 5.5 billion human beings on Earth. All humans belong to the same animal species – *Homo sapiens sapiens*. Like cats, horses and dolphins, humans are mammals, and of all the mammals they are the most successful.

▶ Skin colour is the result of how much melanin (dark colouring) your skin contains, and that depends on which part of the world your ancestors came from. Melanin protects the skin from the Sun's burning rays, so people with dark skin originally came from hot, sunny countries. Pale-skinned people came from colder countries where the sunshine is weaker.

The key to the success of humankind is our intelligence. It has enabled us to use fire, make tools, construct shelters, grow food and clothe our bodies. We have increasingly adapted the environment to our needs. As a result, human beings have spread throughout the world, from forests and deserts to remote islands and the Arctic ice.

Social animals

Human beings are social animals. This means that we grow up in families and choose to live our lives in groups. Since prehistoric times human beings have banded together in tribes and clans to share important tasks such as food-gathering and child-rearing, and to provide security for each other. Our skill with language enables us to share knowledge and ideas, helping to keep the group together.

About 10,000 years ago small human groups discovered how to grow crops and settled down to farm the land. Since then, human settlements have grown from simple villages to vast cities, where millions of people live and work, and from which money, food and goods are traded across the world.

Rich variety

Tall and short, fat and thin, black and white: although human bodies are all built to the same plan, they are all different. Some of this variety is explained by the fact that human beings have adapted to their environment as they have spread. The Inuit people of the Arctic, for example, have relatively short stocky bodies. This shape helps them to keep warm by reducing the surface area through which heat can escape. In contrast, people who live in hot dry climates, such as the Watusi in central Africa, have tall slender bodies, which are less likely to overheat.

As well as physical variety, there is a huge variety in human culture from place to place. The food we eat, the music we listen to and the religion we follow are just some of the things that make up our culture. Today, as people are more mobile, cultures are mixing more and more.

However, when people move from one culture to another they may keep links with their original culture, or 'roots'. *Ethnic groups* are communities of people who feel they have a common cultural identity. They may share a language, a religion and traditional customs. The USA, for example, is often called a multi-ethnic society because many people there would describe themselves as being members of an ethnic group (for instance Jewish, Hispanic or Chinese) at the same time as being American.

▼ For most of our history, humans have been hunter-gatherers. Hunter-gatherers live by hunting animals, fishing and gathering wild foods such as fruits, nuts, vegetables and honey. This way of life began to disappear with the beginning of farming around 10,000 years ago, but still survives among some people, such as the Aeta of the Philippines. This Aeta man is hunting fish with a bow and arrow.

Hunting birds

Hunting birds, or 'birds of prey', such as eagles, falcons and hawks, are superbly equipped to hunt and kill other animals for food. They have hooked bills, and powerful feet with talons. Their sight is better than ours and they have good hearing.

Vultures are powerful birds related to these hunters, but they usually feed by scavenging on the remains of dead animals (carrion). Owls are sometimes called the nocturnal (night-time) birds of prey, but they are not closely related to falcons or vultures.

▶ This red kite soars in the sky hunting for prey. Most birds of prey have such good eyesight that they can spot a meal from a great height.

Out to kill

Hunting methods and types of prey vary enormously. Harriers fly low over the ground searching for small mammals and birds. Sparrowhawks have long legs for grasping prey. They also have short, broad wings and long tails for manoeuvring in the woods where they live. Peregrines use their speed to kill other birds in flight. They can reach speeds of 180 kilometres an hour as they dive on their prey. Lammergeyers drop bones onto rocks from a height to get at the marrow. Honey buzzards that eat bees have blunt talons for digging out bees' nests. Vultures soar, searching for signs of carrion, and rarely hunt live prey. Ospreys have spikes on the soles of their feet to help them grip the fish they catch. Kestrels hover about 10 metres above the ground, looking for small mammals.

Most owls hunt at night. Their large, forward-facing eyes allow them to see in poor light and to judge distance accurately. Owls have very good hearing, so they can hunt even in total darkness. Great grey owls can hear their prey moving under the snow. An owl's soft feathers allow it to fly silently, so its prey does not hear it approaching. Some owls eat small animals, others catch insects, and a few catch fish.

Falconry

Falconry is an ancient sport that uses trained birds of prey to hunt wild birds and other animals. It may have originated in China 4000 years ago. It was common in Europe in the Middle Ages and has recently been revived. Hawks, falcons, eagles and buzzards are the most commonly kept species. Once all birds were taken from the wild. Now this is illegal in many places, and most are bred in captivity.

Hunters hunted

For years humans have shot, trapped and poisoned hunting birds to protect game birds from being preyed on by them. Many have been taken into captivity for falconry. Certain chemical pesticides used in farming are also harmful to hunting birds, because their prey becomes poisoned. The most harmful of these have been banned in many countries. Many populations, such as peregrine falcons, have declined rapidly as a result of chemical poisoning.

Hunting-bird records

Largest
Andean condor.
length 116 cm
Largest owl
Eagle owl: length 70 cm
Smallest
Black-legged falconet: length 14 cm; elf owl: 13–15 cm

• Owls swallow their prey whole and then bring the inedible parts back up as pellets or castings. If you pull these pellets apart, you can see what the owl has eaten by the bones and other food remains in them.

find out more
Birds
Eyes
Food chains and webs

▼ A little owl in flight. To flap their wings, birds have powerful muscles attached to a strong breastbone.

Hyenas

find out more
Grasslands
Mammals

Hyenas are a small group of meat-eating mammals (carnivores). There are four kinds: the spotted, brown and striped hyenas, and the aardwolf.

All four kinds of hyena live in the grasslands and semi-deserts of Africa south of the Sahara. The striped hyena is also found across the Middle East.

Hyenas range in size from the spotted hyena, which has a body up to 1.8 metres long, to the small aardwolf, whose body is no longer than 80 centimetres. The aardwolf feeds mainly on termites, but the other hyenas are mostly scavengers – they eat the remains of animals killed by hunters such as lions and leopards. These hyenas have very powerful jaws and large teeth, and can crush bones and bite through tough skin.

Spotted hyenas are also active hunters. They live and hunt in packs of 10 to 30. Each pack has its own territory. They communicate with each other by smell and a variety of howls, screams, and laughing sounds.

◄ These spotted hyenas in Kenya have just attacked a wildebeest. Spotted hyenas usually bring down their prey by biting its legs. They then tear it to pieces while it is still alive.

Immunity

Immunity is the ability of an animal to defend itself against infection and fight disease. Different parts of your body make up the immune system.

When 'foreign bodies', such as viruses and bacteria, invade your body they can cause disease. The immune system recognizes that chemical groups (molecules) called *antigens* on the surface of these 'invaders' are not part of your body and that the germs that carry them are not wanted. It immediately starts to make *antibodies*, which react with the antigens, making the germs harmless. Once the antibodies against a particular disease have been formed, your body can produce them again when faced with the same 'invaders'. This is called *active immunity* and may be long-lasting. When antibodies from someone who is already immune are injected into you, you have *passive immunity*. This only lasts a number of weeks.

Various parts of the body are involved in the immune system. The thymus helps in the development of the immune system. White blood cells fight 'invaders', and make antibodies. Bone marrow helps make white blood cells. The lymph nodes help get rid of bacteria, and make antibodies and white blood cells. The spleen helps keep the blood clean, and makes antibodies. The tonsils also help to protect you from infection.

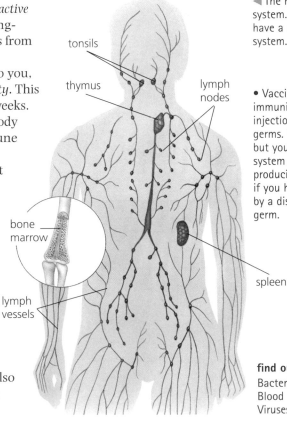

◄ The human immune system. All mammals have a similar immune system.

tonsils
thymus
lymph nodes
bone marrow
lymph vessels
spleen

• Vaccinations or immunizations are injections of harmless germs. You do not get ill, but your body's immune system responds by producing antibodies as if you had been invaded by a disease-causing germ.

find out more
Bacteria
Blood
Viruses

Insects

Together with birds and bats, insects are the only animals that can fly. There are more insects than any other kind of animal. About a million different species have so far been identified, and new ones are being discovered all the time.

An insect's body is divided into three main sections. Its mouth, eyes and antennae (feelers), with which it smells as well as feels, are on its *head*. The *thorax* is concerned with movement: most insects have three pairs of legs and two pairs of wings attached to the thorax. The main part of the gut and digestive system is in the *abdomen*, and so are the sex organs.

▶ Parts of the body of a typical insect.

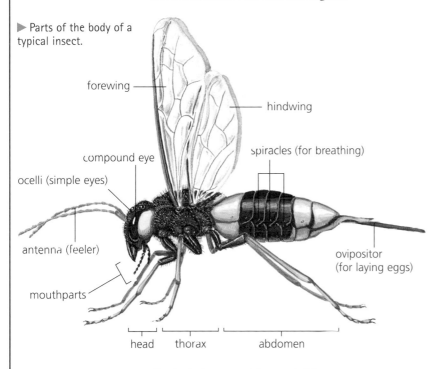

forewing

hindwing

compound eye

spiracles (for breathing)

ocelli (simple eyes)

antenna (feeler)

mouthparts

ovipositor (for laying eggs)

head   thorax   abdomen

• Dragonflies are the fastest-flying of all insects. Some reach speeds of over 50 km/hr. They are very agile in flight and are able to fly backwards.

A skeleton on the outside

Like crabs and spiders, to which they are related, insects have bodies that are supported and protected by a hard outer armour, the *exoskeleton*. This is waterproof and prevents them from drying up in the air. It enables insects to live almost everywhere in the world, even in deserts and on mountains.

The great disadvantage of the exoskeleton is that it cannot stretch, so all insects have to shed their skin (*moult*) in order to grow. As a result, they grow in a series of jumps. After feeding and growing for a time, the hard exoskeleton splits and the insect, which has grown a new flexible skin beneath the old one, wriggles out. At first

this new cover is soft and can expand to allow for growth. But it hardens quite quickly to protect and limit the insect once more.

Insects in the environment

Scientists think that there are probably at least 7 million kinds of insect that we do not yet know about. One reason for the great numbers of different sorts of insect is that they are all small and very many of them can share the same environment. Even a single tree may be home to many insects. Some may be among the roots, some on or under the bark and some on the leaves. Some may be active during the daytime, while others are active at night. Different kinds of insect are active at different times of the year.

Most insects feed on only one type of food, but between them they can eat almost anything. Because of this, some of them are pests of human crops and stores. Most, however, do no damage and are a vital part of the world that we share with them. Bees and butterflies, which pollinate flowers and crops, are useful to us. So are many other insects, such as ladybirds and some hoverflies, which feed on pests such as

▼ These ladybirds are feeding on aphids (greenfly), which are considered pests because they attack crops and spread disease. Aphids reproduce very quickly. During the summer all aphids are female. They do not lay eggs, or even mate, but give birth to live female young, which in turn are able to reproduce in 10 days' time.

Incomplete metamorphosis

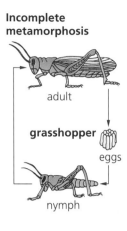

adult

grasshopper

eggs

nymph

Complete metamorphosis

adult

butterfly

pupa

eggs

larva

aphids (greenfly). But we rarely think of the most important insects. These are the recyclers, such as beetles and cockroaches, which feed on the remains of dead plants and animals. As these remains pass through the insects, they are broken down into chemicals which fertilize the soil when returned to it. Plants can then make use of these chemicals when they grow.

The main groups of insects

It is likely that the first insects, which evolved more than 300 million years ago, had no wings, although their bodies were divided into three sections, like the insects of today. A few kinds of *wingless insects* still survive. They are scavengers, and are common wherever there are dead leaves and plants. The young, when they hatch from the egg, look like tiny versions of the adults.

Most insects have wings and can fly. Some, like grasshoppers and bugs, produce young which look like their parents, but at first have no wings. At each moult their wings get bigger. At last, when they are fully grown, they are able to fly. As adults, they feed on the same food as they did before. Insects that develop in this way are said to undergo *incomplete metamorphosis*.

The third group of insects have young that hatch from their eggs as larvae (grubs). They

▲ A shield bug sucks sap from a flower. Bugs are a large and varied group of insects that have one thing in common: they cannot eat solid food. All bugs have long, thin, sharp-tipped mouthparts to pierce the stems, leaves or fruit of plants to suck up the sap, or the skin of animals to drink blood.

look very different from their parents and they feed on different food. When they are fully grown, they turn into pupae, a kind of resting stage during which they change to the adult form. Many of the adults feed chiefly on nectar, and are important as pollinators of flowers. Most of the more familiar insects go through this form of development, known as *complete metamorphosis*.

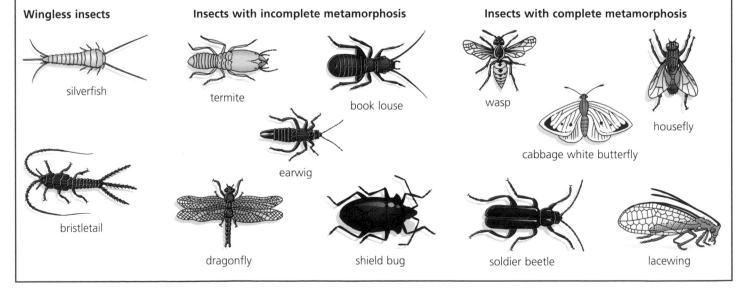

Wingless insects

silverfish

bristletail

Insects with incomplete metamorphosis

termite

book louse

earwig

dragonfly

shield bug

Insects with complete metamorphosis

wasp

cabbage white butterfly

housefly

soldier beetle

lacewing

Jellyfishes and corals

Jellyfishes, corals and sea anemones belong to the same group of animals. They all have a simple tube-shaped or cup-shaped body.

• There are 6500 different kinds of coral and sea anemone, and 200 different kinds of jellyfish.

• Several related but different sorts of creature are often referred to as jellyfishes. These include the Portuguese man-of-war, the sting of which is dangerous to humans.

find out more
Animals
Oceans and seas
Seashore

Much of a jellyfish's bell-shaped body is made up of a stiff, jelly-like material. From the edge of the bell, long tentacles trail into the sea. These tentacles are covered with stinging cells which are powerful enough to paralyse prey such as fishes, shrimps and other sea animals.

Sea anemones and their relatives the corals look like plants but are in fact creatures called polyps. A polyp has a simple, cup-shaped body with a ring of tentacles around the mouth. The tentacles are studded with stinging cells that paralyse and kill any tiny creature that touches them.

Sea anemones are entirely soft-bodied. They live in shallow or deep water, attaching themselves to a hard surface such as a rock, a coral or a shell. If they are taken out of water, they are just like blobs of jelly, but when covered with water they look like flowers, with their frill of short tentacles.

▲ Sea anemones and corals can often be found together. Here a sea anemone lives on a sea-fan coral.

A coral polyp is able to take minerals from the sea water and build them into a hard cup-like skeleton that supports and protects it. Some corals live alone in cold or deep water, but many others are colonial animals and share a huge skeleton. These are the builders of coral reefs. Such reefs are home to many kinds of fishes and other creatures.

Kidneys

As the blood travels round your body, it collects unwanted waste and poisons. Your kidneys' job is to clean your blood and get rid of this waste matter. They do this by filtering the blood and passing the waste out of the body in urine.

• The process of cleaning your blood and getting rid of waste matter is called *excretion*. The body's waste-disposal system is called the *excretory system*.

▶ A kidney cut in half. Kidneys are part of the body's waste-disposal system. They filter blood to stop your body being poisoned by its own wastes.

find out more
Blood

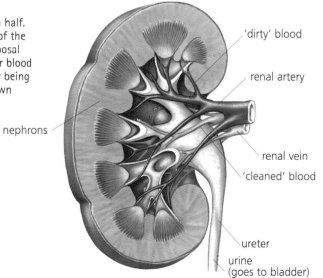

nephrons

'dirty' blood

renal artery

renal vein

'cleaned' blood

ureter

urine
(goes to bladder)

You have two bean-shaped kidneys towards the back of your body, just above the waist. Each one is joined to the bladder by a tube called a ureter. Urine passes down the ureters to the bladder, and from there you pass it out of your body when you urinate.

All vertebrates (animals with a backbone) have kidneys similar to those of a human. In amphibians, reptiles and birds, urine and faeces are expelled through the same passage, known as the *cloaca*, which is also used for reproduction. To save water, reptiles and birds produce nearly solid urine (the white substance in their droppings). In contrast, humans and other mammals lose a lot of water in their urine.

How the kidneys work

Blood enters the kidneys at high pressure and then passes into millions of tiny tubes called nephrons, where waste products are filtered out as urine. The main waste in a mammal's urine is a chemical called urea. This contains nitrogen, which is formed when proteins are broken down in the body. The cleaned blood leaves the kidneys by the renal veins. Your kidneys filter your blood about 50 times a day.

Kangaroos, Koalas *see* Marsupials • **Lemurs** *see* Primates • **Leopards, Lions** *see* Cats

Living things

Human beings are living things, as are whales and mosquitoes, oak trees and mosses, even bacteria. Despite their differences in size and appearance, living things have several things in common: they are able to grow for part or all of their lives; they are able to reproduce (to make a new generation of young); and they are able to react to changes in their surroundings.

This reaction to change can take many different forms. A plant may react to a change in the weather by losing its leaves or opening its flowers. An amoeba (a single-celled animal) might move towards a smell that may mean food. Larger, more complex animals behave in still more complicated ways.

An example of classification

Kingdom Animalia	All animals belong to the kingdom Animalia.
Phylum Chordata	All animals that have a stiffening rod in their backs in the early stages of their lives belong to the phylum Chordata (chordates).
Subphylum Vertebrata	All animals that have backbones belong to the subphylum Vertebrata (vertebrates).
Class Mammalia	All animals that breathe air, are hairy and feed their young on milk belong to the class Mammalia (mammals).
Order Perissodactyla	All animals that are hoofed and have an odd number of toes – including horses and their relatives – belong to the order Perissodactyla.
Family Equidae	All horse-like animals, from about 60 million years ago to the present day, belong to the family Equidae.
Genus Equus	All single-toed horses, from about 5 million years ago – including horses, donkeys and zebras – belong to the genus *Equus*.
Species Equus caballus	The domestic horse belongs to the *caballus* species. The scientific name for any animal species combines its genus and species name, so the domestic horse is properly known as *Equus caballus*.

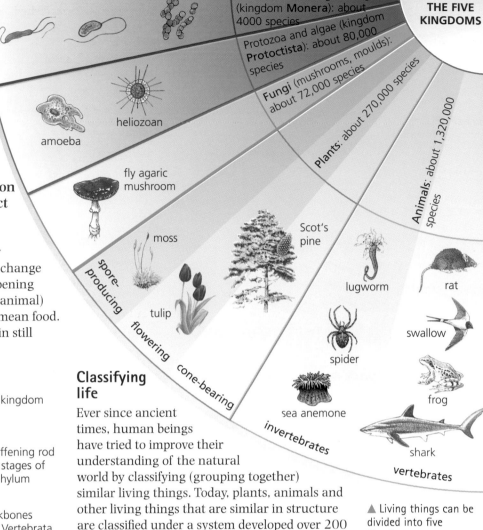

Bacteria and blue-green algae (kingdom **Monera**): about 4000 species

Protozoa and algae (kingdom **Protoctista**): about 80,000 species

Fungi (mushrooms, moulds): about 72,000 species

Plants: about 270,000 species

Animals: about 1,320,000 species

THE FIVE KINGDOMS

heliozoan
amoeba
fly agaric mushroom
moss
tulip
Scot's pine
lugworm
rat
spider
swallow
sea anemone
frog
shark

spore-producing
flowering
cone-bearing
invertebrates
vertebrates

▲ Living things can be divided into five kingdoms.

Classifying life

Ever since ancient times, human beings have tried to improve their understanding of the natural world by classifying (grouping together) similar living things. Today, plants, animals and other living things that are similar in structure are classified under a system developed over 200 years ago by a Swede, Carolus Linnaeus.

The basic unit of the Linnaean system is the individual. Although no two individuals are exactly the same, some are clearly very alike. They belong to the same *species*, a group that is able to breed successfully in the wild. Similar species belong to the same *genus* (plural: genera). They cannot breed together. Genera are grouped into *families*, which consist of clearly related animals or plants. For instance, all cats belong to one family; all dogs to another.

Families are put together into *orders*. Cats and dogs both belong to the same order, along with bears and weasels among other flesh-eaters. Orders are grouped into *classes*. All mammals belong to one class. A number of classes together make up a *phylum* (plural: phyla). A phylum includes animals or plants with basic similarities, though these are often masked by adaptations to a particular way of life. Finally, the phyla are grouped into five *kingdoms* – animals, plants, fungi, Protoctista and Monera.

• More than 1.75 million different species of living thing have been identified, and more are being discovered all the time. The total number could be in the region of 13 million! The numbers of identified species belonging to the Monera and Protoctista kingdoms are probably only a tiny proportion of the total numbers that exist.

find out more
Animals
Bacteria
Fungi
Plants
Viruses

Lizards

Lizards are small, agile reptiles. Most live in warmer, drier parts of the world, and can often be seen sunning themselves or scuttling away from danger.

All lizards have sharp eyesight and a good sense of hearing. They generally feed on insects and other small animals. Some are burrowers, hunting their prey underground. Lizards' skin is silky smooth and their legs are small. Some, like slow-worms, have no legs at all.

Lizards have many enemies. They protect themselves by moving quickly and by camouflage – they match their environment and so are difficult to see.

Some types of lizard shed their tails if caught by a predator, and so escape. The tail will grow again.

Chameleons are small lizards that have a very special kind of camouflage. They change colour, sometimes in as little as two minutes. In bright sunshine they are a rich green, while at night they become whiter. A frightened chameleon may turn pale, while an angry one goes blackish green.

Generally, lizards avoid people, but geckos like to live in buildings. They have suction pads on their feet which help them to climb walls or walk upside-down on ceilings. They are welcome in many houses, because they feed on flies and other household pests.

Most lizards lay eggs, but some kinds give birth to live young, and a few even feed developing young through a placenta like mammals.

common wall lizard

blue-tailed day gecko

chameleon

• There are about 3000 different kinds of lizard, more than any other kind of reptile. The largest lizard is the Komodo dragon, which grows to over 3 m long and weighs over 100 kg.

find out more
Reptiles

Lungs

Humans and many other animals have organs called lungs which are used for respiration, the process by which the body gets the energy it needs to function. The lungs absorb oxygen from the air breathed in, and extract the waste gas carbon dioxide from the blood after respiration.

Air gets to and from the lungs via the windpipe (trachea) which leads from the throat. The trachea divides into two main air tubes called *bronchi*, one going to each of the two lungs. The bronchi continue to branch and divide into thousands of very tiny air tubes called *bronchioles*. At the end of each bronchiole is a group of tiny air-filled sacs called *alveoli*.

The oxygen from the air breathed in passes from these alveoli into tiny blood vessels (capillaries) nearby. It then travels around the body to special cells where respiration takes place. Respiration uses oxygen to release energy from food, and it produces water and carbon dioxide as by-products. Carbon dioxide is not needed by the body, so it is carried in blood vessels back to the lungs. It passes through the alveoli into the bronchioles and is breathed out when you exhale.

▶ Air flows in through the nose and mouth, down the windpipe and into the lungs via the two bronchi. The lungs are like a soft pink sponge because they are made up of millions of tiny air tubes (bronchioles) which end in air sacs (alveoli) covered with blood vessels.

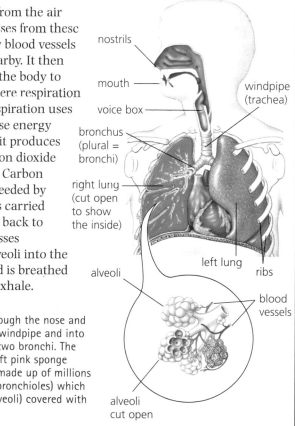

nostrils

mouth

voice box

windpipe (trachea)

bronchus (plural = bronchi)

right lung (cut open to show the inside)

left lung

ribs

alveoli

blood vessels

alveoli cut open

• All mammals, birds, reptiles and air-breathing amphibians have lungs. Most fishes do not have lungs. However, one group that does is called the lungfishes. Their lungs are filled with air when they put their noses through the water surface. They also have gills which they use underwater.

find out more
Blood
Breathing

Mammals

Mammals are the creatures that most of us think of when we use the word 'animals'. We know them best among all living things, because many of our domestic animals are mammals and so are we. In general, mammals are far more intelligent than other animals.

Mammals are found almost everywhere in the world, from the cold lands and seas of the Arctic to hot deserts and steamy forests. There are about 4000 different kinds of mammal. Some are plant-eating, some flesh-eating, and some are omnivores – they eat both plants and flesh. Bats can fly; whales live only in water; antelopes and horses are runners; monkeys are climbers.

What is it about these animals, different in size, appearance and way of life, that makes us group them all together as mammals?

They have five things in common:
- All mammals have bones, including a backbone.
- All mammals have lungs and breathe air.
- All mammals are warm-blooded.
- All mammals have some fur or hair on their bodies at some stage in their life.
- All female mammals feed their babies on milk.

The first mammals were small and evolved at about the same time as the dinosaurs, 220 million years ago. When dinosaurs died out 65 million years ago, the numbers of mammals increased dramatically, and they quickly developed into shapes and sizes that suited them for life in almost every habitat. Mammals became the dominant land animals, and the biggest animals on land and in the sea. Almost all large mammals are now under threat of extinction (dying out), mainly because of the activities of human beings.

▶ Duck-billed platypus.

▲ A young orang-utan suckles at its mother's breast. The name mammal comes from the Latin word for breast, *mamma*, as all female mammals produce milk to feed their young.

Mammal groups

Mammals are divided into three groups depending on how their young develop.

The monotremes

Monotremes are the only mammals that hatch from eggs. There are only three kinds of monotreme, and they all live in Australia or New Guinea: two kinds of echidna, or spiny anteater, and the duck-billed platypus. When the young platypuses hatch from their eggs, they lap up milk from tiny glands on their mother's belly.

The marsupials

Baby marsupials are minute when they are born – some are no larger than a grain of rice – but all depend on their mother's milk to complete their growth. In their mother's pouch they feed on milk and grow in safety. Marsupials are found mainly in Australia and New Guinea, but some also live in South and North America.

The placentals

Most mammals, including whales, bats, giraffes, hippopotamuses and human beings, are placentals. Young placentals grow inside their mother's body for a long while before they are born. They are nourished through a special organ called the placenta, which develops within the mother's body.

Mammal records
Largest
Blue whale, up to 33 m in length and about 120 tonnes in weight
Smallest
Several kinds of shrew have a head-and-body length of less than 4.5 cm and weigh about 2 g. The hog-nosed bat of Thailand is about the same weight.
Tallest
Giraffe, up to 5.3 m

find out more
Animals
Growth and development
Marsupials
Prehistoric life
Sex and reproduction

Marsupials

The group of animals known as marsupials includes kangaroos, koalas and opossums. They are found only in Australia, New Guinea, and North and South America. Most female marsupials have a furry pouch on their bellies. This pouch holds and protects the young from the time they are born until they are big enough to fend for themselves.

When baby marsupials are born, they have hardly begun to develop and are very tiny. Some are as small as a grain of rice. But they have strong forelimbs and claws, which they use to crawl through their mother's fur to her pouch. There they find a teat, and fasten themselves to their food supply. Baby marsupials grow fairly slowly.

Marsupial records
Largest
Red kangaroo: males may measure 2.5 m total length, and weigh up to 90 kg.
Smallest
Pilbara ningaui: head-and-body length of adult may be as little as 4.6 cm; weight may be no more than 2 g.

• There are about 280 different kinds of marsupial.

▶ A Tasmanian pademelon with its young feeding in a eucalyptus forest. Pademelons, or scrub wallabies, are hunted for meat and fur.

Where marsupials live

Marsupials are found mainly in Australia, New Guinea and South America. This is because for much of the last 100 million years Australia and South America have been island continents. When they became separated from the other continents, almost the only mammals living in Australia were marsupials, and in South America some mammals were marsupial. In both continents marsupials evolved into many different sorts of animal. When South America became joined to North America about 10 million years ago, most of the South American marsupials were killed off by more successful mammals which invaded from the north.

• The Virginia opossum pretends to be dead if it is caught, or badly frightened. This is known as 'playing possum'. Often its enemy loses interest and the opossum makes its escape.

find out more
Mammals
Prehistoric life

Kangaroos, koalas and wombats

The kangaroos of Australia and New Guinea, and their smaller relatives the wallabies and rat kangaroos, have powerful back legs. When moving slowly, they drop onto all fours. At higher speeds, they hop. The biggest kangaroos can cover more than 10 metres in a single bound. All kangaroos are plant-eaters, active mainly at dusk or at night. Several smaller kinds of kangaroo are endangered.

Koalas are climbing animals. They live among the branches of eucalyptus forests in eastern Australia, and feed almost entirely on the young leaves and shoots of the eucalyptus. Koalas generally sleep for about 18 hours a day. In the past they have been hunted for their fur, but they are now protected.

The wombats of Australia live on the ground and burrow. A common wombat has several burrows which it may share with other wombats. Wombats are active at night, when they emerge to feed on grasses and roots.

Opossums

Opossums live in North and South America. Some species have pouches, but many have two folds of skin on their bellies, or no pouch at all. Most opossums live in forests and have hairless tails that they use to cling to branches. They feed mainly on small animals, including insects.

▶ Marsupials vary greatly in appearance and in how they live. There are grazing, climbing, burrowing, and flesh-eating marsupials. The Tasmanian wolf has not been seen since 1936, so it is almost certainly extinct.

kangaroo

koala

wombat

cuscus

Tasmanian devil

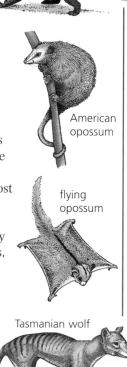

American opossum

flying opossum

Tasmanian wolf

Mice, squirrels and other rodents

Mice, rats, gerbils, hamsters, guinea pigs, squirrels, beavers and porcupines all belong to the group of animals known as rodents. About half of all of the known kinds of mammals are rodents, and they are found in almost all parts of the world, from hot tropical forests to deserts and cold tundra. Many other mammals and birds feed on rodents, so most kinds of rodent produce enormous numbers of young so that some will survive.

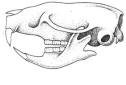

▼ A skull of a rat, showing the razor-sharp incisor teeth at the front of the jaw and the grinding teeth at the back.

The name 'rodent' comes from the Latin word meaning 'to gnaw'. At the front of their mouths rodents have only two razor-sharp incisor teeth in the upper jaw and two in the lower jaw. At the back of their mouths are grinding teeth, because rodents feed mainly on plants, though many eat insects and other small creatures as well.

Mice and dormice

Different kinds of mice are found throughout most of the world. They live in almost all habitats, from the harshest deserts to lush tropical rainforest. Most mice have little contact with human beings. Only one, the house mouse, has become a common pest, mostly because of the damage it causes.

wood mouse

Dormice are small, plump, furry-tailed rodents with long whiskers and large eyes. They are good climbers and usually live in woodlands or rocky places. The word 'dormouse' means 'sleep mouse', and most dormice hibernate (go into a deep sleep) for more than six months in winter.

Rats

The most common rats are the black rat and the brown rat. Both have become major pests to humans. They eat almost anything, and both can carry diseases, some of which may be passed to human beings or domestic animals. The Black Death, which is said to have killed a third of the population of Europe in the Middle Ages, was caused by germs carried by the fleas of black rats.

Black rats need warmth and so they live mostly in countries with hot climates. Brown rats are more hardy and can be found in almost all parts of the world, partly because humans have brought them on board ship by mistake.

• Brown rats have been tamed and make excellent pets, despite the bad reputation of their wild relatives. They are very clean and much friendlier than the more commonly kept rodents, such as hamsters and gerbils.

Voles and lemmings

Voles are sometimes mistaken for mice, but they have shorter tails and legs, and blunt faces with small eyes and ears. They are most often found in fields and meadows, though some live in woodland or desert areas. They rarely live for more than a year, but females can produce several litters of young in this time. Many of the young are eaten by foxes, owls or other predators.

Some voles are known as lemmings. Lemmings are found in burrows or rock crevices in northern tundra regions of America, Europe and Asia. About once every four years the number of lemmings increases dramatically. Many move away from where they were born to find new sources of food. In their hurry, some may rush into rivers or even the sea, where they drown. This has caused people to believe, mistakenly, that lemmings commit suicide.

lemming

Gerbils, hamsters and guinea pigs

In the wild, almost all true gerbils live on the edge of deserts in Africa and the Middle East. They escape from the daytime heat in deep burrows and come out at night to feed. There are many kinds of gerbil and many kinds of jird and sand-rat, their close relatives. Often people keep jirds as pets, but call them gerbils by mistake. Jirds have shorter hind legs than gerbils and are a bit rounder and fatter.

Hamsters are plump-bodied rodents. Most have short faces, and short legs and tails. They live in burrows in the fields of Europe and western Asia. All hamsters have huge pouches at the side of their mouths. These are used like built-in shopping baskets to carry food. Golden hamsters are popular as pets.

Guinea pigs are small tail-less rodents. They got their name because when they were first brought to Europe from South America in the 16th century, people thought they looked like tiny pigs.

eastern chipmunk

Squirrels

Squirrels are found in almost all parts of the world. Unlike most small mammals, they are active during the day. Like many rodents, though, squirrels store extra food, usually by burying it.

Tree squirrels, such as the grey and red squirrels, are well adapted for tree-climbing. Their sharp claws help them to cling onto branches, while their long fluffy tail helps them to keep their balance. *Flying squirrels* have membranes (thin layers of skin) between their legs that enable them to glide from tree to tree. Other types of squirrels are known as *ground squirrels*, and include prairie dogs, woodchucks (or groundhogs), chipmunks and marmots.

Porcupines and beavers

Porcupines are large rodents with sharp spines or quills over most of their bodies, which they use for defence. There are two main groups of porcupines. New World (American) porcupines live mainly in trees. They have flexible tails, which help them to climb and keep their balance, and they generally have short quills. Old World porcupines live in warm areas, from southern Europe to Borneo and Malaya. Most kinds live on the ground and have long black-and-white quills.

Beavers are water-living rodents found in northern parts of Europe, Asia and North America. More than any other animal, except humans, they are able to change their environment. They build dams across streams by dragging stones, mud and logs from trees they have cut down with their sharp teeth. Usually the dam contains a lodge – a space in which they live and store food.

Because beaver fur is valuable, the number of beavers in the world is low and they are specially protected.

grey squirrel

Old World porcupine

◀ A beaver's lodge, made of branches and compacted mud, can be as tall as a human, although most of it is underwater. The central chamber is lined with twigs and in spring two to eight young are born here. Part of the roof is made only of sticks, so that some air can get in.

find out more
Animals
Ecology
Hibernation
Mammals

Migration

Migration is the regular movement of animals to and from a particular area. The most common type of migration is seasonal, when animals travel from one area to another at different times of the year.

The journeys made by migrants are often very long. The Arctic tern, for example, travels from the pack ice of the Antarctic to the most northerly parts of Europe, Asia and North America to breed. By contrast, some kinds of hummingbird move only a few hundred metres up and down a mountain slope as the seasons change.

Migration may also take place on a longer time scale. Salmon spend the first part of their lives in rivers, migrate to the ocean as adults, and then return to the rivers to breed. Eels do the opposite – they hatch in the sea and migrate to fresh water to grow. A few kinds of animals are on the move almost constantly. Wildebeest in Africa and, in the oceans, some kinds of whales make a circular tour in the course of a year.

In winter some birds and other animals move away from their breeding areas to places where the weather is warmer. Migrants in northern Europe include starlings, some kinds of bat, and reindeer (caribou).

Human migration

People also migrate. Many human migrants move from one country to another for good. However, some peoples, called *nomads*, are always on the move, herding their animals from one place to another in search of fresh grass.

find out more
Animal behaviour
Animals
Birds
Eels
Salmon and trout
Tundra
Turtles and tortoises

▼ Some green turtles migrate over 2000 km, from their feeding grounds on the coast of Brazil to Ascension Island in the Atlantic Ocean, to breed. This strange migration probably evolved over millions of years.

Mongooses

Mongooses are small, short-legged mammals with pointed faces and long bushy tails. They are found in the warmer parts of Europe, Asia and Africa.

Mongooses are fast-moving, agile hunters. They are a very widespread and successful group of animals. None of the 31 different kinds is in danger of extinction.

Mongooses feed mainly on small animals, from insects to birds and small mammals. They also eat birds' eggs and even fruits. They are daring hunters and are best known for their ability to kill snakes, including poisonous ones. They are not immune to the poison, but are so quick that they can avoid the snake's strike.

Most mongooses are solitary creatures, living either alone or in pairs. Some live in family groups, and a few, like the dwarf mongoose and the slender-tailed meerkat, live in larger groups, sometimes numbering up to 40 animals. Females have between two and four young at a time.

In the late 1800s mongooses were introduced into Hawaii and parts of the West Indies. This move was intended to help destroy the local snake and rat populations. However, it was not a success because the mongooses killed local chickens and also attacked the native bird population.

• The largest mongoose, the white-tailed mongoose, grows to a length of up to 58 cm, not including the tail. This is more than double the length of the tiny dwarf mongoose, the smallest of all the mongoose family.

◄ A mongoose will grab a snake by its neck. Once this happens, the fight is over and the snake is eaten by the mongoose.

find out more
Mammals
Snakes

Monkeys

Monkeys are primates, as are human beings, but it is not true to say that we are descended from them. However, monkeys are like us in many ways, particularly the way in which they use their hands for holding things.

▼ The golden marmoset has a reddish-gold coat. Unlike most other monkeys, marmosets have claws rather than nails on their hands and feet.

Monkeys have very good eyesight, which is important as many of them spend a lot of their time leaping between branches. Like humans, most monkeys have five fingers and toes on their hands and feet, and nails rather than claws.

Monkeys are sociable animals, living within groups of their own kind. Within the group each monkey knows its place, and the leader is rarely challenged. There may be noisy arguments and occasional fights, but group members are unlikely to hurt each other seriously.

Female monkeys usually give birth to one baby every few years. These are mostly carried by their mother, at first clinging to her underside and later riding on her back. Young monkeys develop slowly, and they do not usually become adults for several years.

New and Old World monkeys

There are two main groups of monkeys. Those that live in South America – the New World monkeys – all live in trees. They are wonderful climbers, and many have prehensile (gripping) tails which they use like an extra hand. New World monkeys include spider monkeys, howler monkeys, marmosets and tamarins.

The other kind of monkeys – the Old World monkeys – are found in forested areas of Africa and the warm parts of Asia. None of these has a prehensile tail, although most have some kind of tail. Some, like baboons, live mainly on the ground. Others, such as the colobus monkeys, are expert climbers and rarely leave the tree-tops.

When they are resting, all of the Old World monkeys sit upright. They have areas of hardened skin on their

buttocks, which act as built-in cushions. None of the New World monkeys have this patch.

Monkeys at risk

Some monkeys have become pests, raiding crops from farms and taking food left out in open places. In India, monkeys are sacred animals and are protected, but in many parts of the world they are now becoming very rare. This is partly because they are hunted for food, for their skins, and sometimes for use in laboratories, but mostly because the forests in which they live are being destroyed.

▶ The spider monkey has a strong, prehensile (gripping) tail that has a sensitive tip. It can be used like an extra hand to feel for food.

◀ The mandrill is a kind of baboon. The coat of the male is dark but he has a bright red nose and bright blue cheeks. He also has a blue-red patch on his buttocks. Males have these markings to help them attract females.

▼ Colobus monkeys are either black and white – like this one – or red. Many thousands used to be killed because their coats were much in demand for making clothes.

• The New World monkeys from South America have flat noses with nostrils set wide apart and opening to the side. The nostrils of the Old World monkeys from Africa and Asia are much closer together.

find out more
Apes
Human beings
Mammals
Primates

Moors and heaths

Moors and heaths are open treeless areas. Some are found where it is too wet or windy, or the soil is too poor and peaty, for trees to grow. Others were formed when people cleared the trees for hunting or to graze sheep or cattle.

Heaths are covered in low-growing heather-like shrubs, while moors are made up of coarse grasses and sedges. If grazing is very heavy, heaths turn into grassy moors. Heaths can become very dry in summer, and fires are common, caused either by lightning or by humans. Without fires or grazing, heaths would soon return to woodland.

◄ Moors and heaths are generally bare of trees, but occasional rowan trees (also called mountain ash) are found on this moorland in Lancashire, UK.

● Heaths and moors are full of bird life. Overhead, birds of prey such as buzzards, kestrels and peregrines hunt for birds and small mammals.

find out more
Grasslands
Wetlands

Muscles

Muscles are parts of an animal's body that make the other parts move. They are needed for all kinds of movement, including walking, running, lifting things, blinking, pumping blood and swallowing food.

● All muscles need energy to work. Blood carries oxygen and glucose (a sugar made from digested food) to the muscles. Through the process of respiration, the muscles get the energy they need to work properly.

Muscles are made up of muscle cells, and are controlled by nerve signals from the brain. They work by contracting (getting shorter). Muscles cannot lengthen themselves, but they can be stretched when other muscles pull at them.

There are about 650 muscles in the human body, and there are three types. The first type is attached to the bones and is used to move the body. It only works when you want it to, so it is called a *voluntary muscle*. When you want to move, it pulls against the skeleton, bending it at joints such as the elbow and the knee.

The second type of muscle can contract without you thinking about it, so it is called an *involuntary muscle*. An example of an involuntary muscle is the muscle around the gut which pushes food along it.

The third type of muscle is found in the heart. This is the *cardiac muscle* which pumps blood around the body. It can work continuously for years without getting tired.

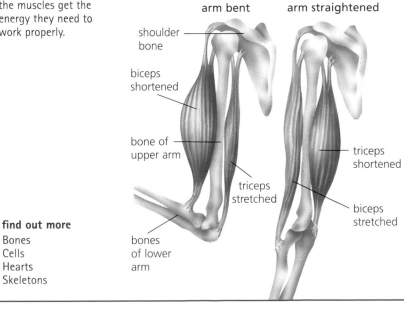

arm bent

arm straightened

shoulder bone

biceps shortened

bone of upper arm

triceps stretched

bones of lower arm

triceps shortened

biceps stretched

find out more
Bones
Cells
Hearts
Skeletons

◄ There are at least two muscles at each joint. In the upper arm of humans the biceps and triceps muscles work together to bend and straighten the arm.

Mussels and oysters

▲ A common mussel. A large mussel filters about 1.5 litres of water an hour to gets its food.

• There are over 20,000 kinds of bivalve mollusc. The other important groups of molluscs are gastropods (slugs, snails, limpets) and cephalods (squids and octopuses).

find out more
Animals
Oceans and seas
Seashore
Slugs and snails
Squids and octopuses

Mussels and oysters belong to a large group of animals called bivalve molluscs. These animals have two shells, joined by a hinge of elastic tissue. Some bivalves live in rivers and lakes, but most are found in the sea.

Most bivalves, such as cockles and razorshells, bury themselves in soft mud or sand, using a muscular 'foot' to burrow their way down.

▶ There are about 1000 kinds of freshwater mussel known. They live in streams, lakes and ponds in most parts of the world.

To feed, they push a long tube up to the surface and suck in water containing plankton (tiny floating plants and animals). The water also holds oxygen, so bivalves breathe and feed at the same time.

Mussels and oysters are unusual in that they live attached to rocks. Mussels are attached with strong threads that hold them safe from the crashing waves. They are sometimes left above sea level as the tide falls, but they do not usually spend very long out of water. Most oysters cement themselves to rocks, usually below low-tide level.

Like other bivalves, mussels and oysters feed on plankton, which they suck into their

▶ One side of a true oyster shell is flat and the other side, in which the body lies, is rounded. Oysters anchor the rounded side of the shell to the seabed and use the flat side as a lid. If danger threatens, they snap the lid firmly shut.

shells with water. Sometimes, by accident, a grain of sand gets inside the shell in a place where it cannot be swept away by a current of water. When this happens, the creature covers the grit with layers of scaly material (mother-of-pearl) to get rid of the irritation. If a perfectly round pearl is formed, it may be very valuable, as humans turn such pearls into expensive jewellery.

Nervous systems

• Reflexes are built into your nervous system. They are actions which you do without having to think, such as pulling your hand away if you touch something hot.

• Pain is an unpleasant feeling that warns us of damage to our bodies. Nerve endings in the skin and other parts of the body carry the pain message to the spinal cord and into the brain.

find out more
Brains
Cells
Glands
Skeletons

The nervous system of humans and other vertebrates (animals with backbones) is made up of the brain, the spinal cord and a network of nerves. It is the body's communication and control system.

The nervous system ensures that all the parts of an animal's body are co-ordinated and that the animal responds to any changes in its surroundings.

The brain and the spinal cord make up the *central nervous system*. This is protected by bone: the brain by the skull and the spinal cord by the backbone. The *peripheral nervous system* is made up of the nerves that connect all the parts of the

body with the brain and the spinal cord.

The nerves carry electrical messages (nerve impulses). Nerve impulses only travel in one direction, which is why we need two different sets of nerves: sensory and motor nerves. *Sensory nerves* carry

information from the sense organs, such as the eyes, ears, nose, taste buds and touch sensors in the skin, to the brain. The brain sends out nerve impulses through the *motor nerves* to the part of the body where action is needed, such as a muscle or a gland.

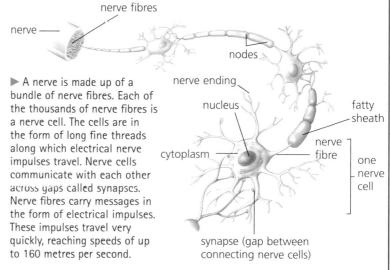

▶ A nerve is made up of a bundle of nerve fibres. Each of the thousands of nerve fibres is a nerve cell. The cells are in the form of long fine threads along which electrical nerve impulses travel. Nerve cells communicate with each other across gaps called synapses. Nerve fibres carry messages in the form of electrical impulses. These impulses travel very quickly, reaching speeds of up to 160 metres per second.

nerve fibres
nerve
nodes
nerve ending
nucleus
fatty sheath
cytoplasm
nerve fibre
one nerve cell
synapse (gap between connecting nerve cells)

Newts and salamanders

Newts and salamanders are amphibians, like frogs and toads. They have long narrow bodies, long tails and short legs with webbed feet for swimming.

They use their powerful tails to propel themselves through the water. Like frogs, newts and salamanders have moist skins and live in damp places. They feed on worms and insects, and some use their tongues to catch their prey. Most newts and salamanders spend their adult lives on land, but some live mainly in water.

Some newts and salamanders lay their eggs in water. These eggs hatch into tadpoles, which feed on small water animals. Others lay their eggs on land, in damp hollows under old logs or stones. The tadpole stage takes place inside the eggs, and the tiny young eventually hatch out from the eggs. A few kinds give birth to live young. A few salamander tadpoles never really grow up. An adult axolotl is like a giant tadpole, still with feathery gills on the outside of its head.

Most newts and salamanders produce poisonous slime from glands in their skin. They may be brightly coloured to warn enemies of their poison.

▼ The fire salamander has bright yellow markings on its black body, tail and head, to warn enemies of its poisonous skin. It spends most of its time on land, living in damp places.

▶ Like most newts, the marbled newt lives mainly on land but returns to water to breed.

find out more
Amphibians

Noses

find out more
Elephants
Senses
Whales and dolphins

Most fishes, amphibians, reptiles, birds and mammals have two nostrils. These are tubes running from the front of the head into the mouth. In most mammals these tubes end in a special bulge on the face called a nose.

The nose and its tubes have two important functions. First, they are a route for air to enter the lungs when the mouth is shut or filled with food. Second, the walls of the nostrils are lined with special nerve cells that respond to odours in the air. These send messages to the brain, and this produces the sense of smell. Humans and other animals also use their nose when they eat, as many food 'tastes' are actually detected by the nose.

Human beings can distinguish between 10,000 and 40,000 different odours. However, many animals have a much better sense of smell. A bloodhound, for example, has a sense of smell that is around one million times better than that of a human, which is why they are used by the police to find hidden drugs or to search for clues that will help solve a crime.

Noses come in all shapes and sizes. An elephant's nose, together with its upper lip, forms the trunk. A whale's nose is the blow-hole on top of its head.

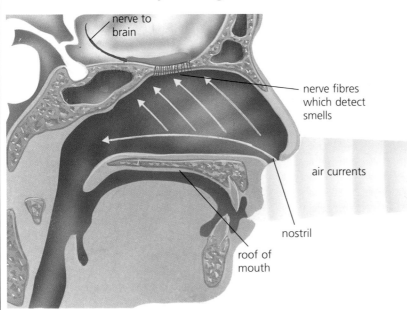

nerve to brain

nerve fibres which detect smells

air currents

nostril

roof of mouth

◀ Chemicals in the air you breathe dissolve in moisture-covered nerve ends in your nose. These chemicals make the nerves send messages to your brain, so producing your sense of smell. Many animals smell in the same way.

Oceans and seas

Oceans and seas cover 71 per cent (over 360 million square kilometres) of the Earth's surface. They contain about 1370 million cubic kilometres of water.

Oceans and seas are also home to a vast number of living creatures, from microscopic plants and animals to giant whales.

Ocean zones

The oceans can be divided into three zones. The sunny surface waters at the top – the *photosynthetic zone* – contain most of the ocean fishes as well as a floating community of billions of microscopic creatures called plankton. Below this zone lie the more dimly lit *twilight zone* and, reaching down to deep cold waters, the *dark zone*. Fewer life forms, mainly flesh-eating fishes, live in the lower zones.

Most of the ocean is at around the same temperature – about 4 °C. As you go down deeper, the pressure of the water above increases steadily, making it difficult to move

▶ The oceans can be divided into three zones. The photosynthetic zone at the top receives enough sunlight to photosynthesize and produce food. It contains most of the ocean's fishes as well as plant and animal plankton. The twilight and dark zones beneath are home to fewer life forms, mainly carnivorous fishes. (This diagram is not to scale.)

quickly. The temperature also falls to around 2 °C in deep water. The amount of light decreases, until at 1000 metres there is no light at all.

Life at the top

The plant and animal plankton in the photosynthetic zone provides food for tiny animals, such as shrimps, prawns, and the young of starfishes, crabs and other sea animals. Away from the sheltered coastal waters there are fewer different kinds of animal, but many fishes and a few large mammals like whales, dolphins and porpoises live here. Some, such as baleen whales and basking sharks, feed by filtering the water for plankton. Others, such as white sharks and barracuda, hunt other fishes. ▶

▼ The world's oceans. They contain almost 97% of all the water found on Earth.

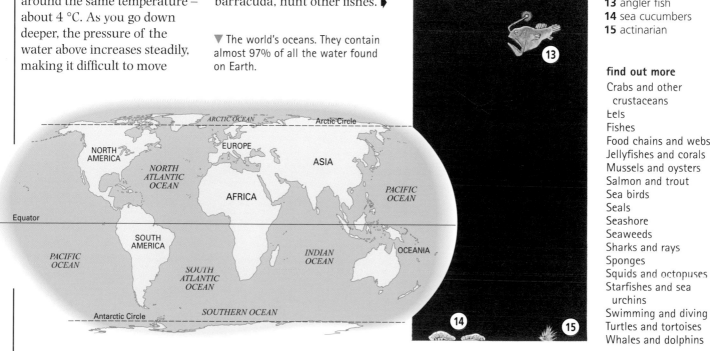

photosynthetic zone (up to 200 m deep)

twilight zone (up to 1000 m deep)

dark zone (the average sea depth is 4000 m, but it reaches 11,000 m at its deepest point)

1 gulls
2 dolphin
3 jellyfish
4 tuna
5 plant plankton
6 turtle
7 animal plankton
8 sperm whale
9 hatchet fish
10 squid
11 rat-tail
12 gulper eel
13 angler fish
14 sea cucumbers
15 actinarian

find out more
Crabs and other crustaceans
Eels
Fishes
Food chains and webs
Jellyfishes and corals
Mussels and oysters
Salmon and trout
Sea birds
Seals
Seashore
Seaweeds
Sharks and rays
Sponges
Squids and octopuses
Starfishes and sea urchins
Swimming and diving
Turtles and tortoises
Whales and dolphins

Deep-sea life

In the cold, dark waters of the deep ocean, hunters can spot the silhouettes of their prey against the faint light above. Here, many fishes have silvery scales along their sides to reflect any light and disguise their shapes. Others are flat-sided, giving them very narrow silhouettes.

In the depths of the ocean food is scarce and meals may be months apart. Many fishes have huge mouths and can eat prey that is larger than themselves. Gulper eels and hatchet fishes swim with their large mouths open to catch whatever they can. Some fishes have elastic stomachs.

▲ A West Indian manatee. Manatees and dugongs are large mammals that live in coastal waters in some warm parts of the world. They can grow up to 3.5 m in length. Some people think that in the past sailors may have mistaken them for mermaids!

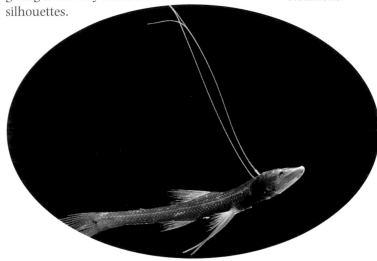

◀ Many bottom-dwelling fishes rely on smell and taste. The tripod fish uses three fins to prop itself above the sea-bed while smelling for prey.

Ostriches

The ostrich is the largest living bird. It can grow to 2.5 metres tall, and weighs up to 150 kilograms. Ostriches live in the grasslands of Africa, where they feeds on plants, but they sometimes also eat small reptiles.

Although their wings are small and useless for flying, ostriches have long, powerful legs, and can outrun most enemies. They can run at speeds up to 70 kilometres per hour, making them the fastest animals on two legs. Their long necks also help them to see when danger approaches.

The ostrich belongs to a group of mostly large, flightless birds that are known as the *ratites*. They include the rheas of South America, the emu and cassowaries of Australia and New Guinea, and the much smaller kiwis of New Zealand. New Zealand was also home to the moas, giant ratites that stood over 3 metres tall, which became extinct 500 years ago.

find out more
Birds
Conservation

▶ A male ostrich with chicks. Male ostriches mate with a number of females, and share in the task of looking after the young.

● An ostrich's egg is about the size of a small coconut. It is about 15 cm long, weighs around 1.35 kg, and holds about 40 times as much as a hen's egg.

Pests and parasites

find out more
Fleas and lice
Flies and mosquitoes
Slugs and snails
Worms

Any living thing that harms humans or human activities can be called a pest. A parasite is a living thing that steals food and shelter from another animal or plant, but gives nothing in return.

▼ These fish lice are parasites of fishes.

In some cases, parasites can also be pests. The varroa mite, which is a parasite on honey bees, is also a serious pest as it weakens the entire bee colony. Infections of varroa have affected honey production in many countries.

Pests

Most pests affect our food in some way, either by eating food crops or by causing disease. Some bacteria and fungi cause diseases in food plants. Insect pests include mosquitoes, which spread disease among animals by drinking their blood, and aphids, which weaken crops by feeding on their sap. Snails and slugs are pests in the garden, and rats and mice can be pests in the house.

Plant pests are often controlled by the use of chemicals called pesticides, which kill the pest but not the plant. Pesticides can cause damage to the environment, so farmers may also use other types of pest control. They may grow crops that are resistant to some pests, or release animals that feed on the pests.

Parasites

The animal or plant that a parasite lives on is called its *host*. Animal parasites like fleas and lice live on the skin of their hosts. They are called *external parasites*. Other parasites, such as tapeworms and hookworms, are *internal parasites*, that live inside their hosts.

A parasite of a large or long-lived host does not generally kill it, but can weaken it and may cause disease. Parasites of small creatures such as insects are often of a similar size to the host, which they usually kill.

Pigs

Pigs are mammals that have short legs, heavy bodies, short tails (which are often curly), and a snout for a nose. Most of them have a coarse, bristly coat, although many farm pigs have a smooth skin.

domestic pig

Pigs are common as domestic animals that are kept on farms and then killed for their skin and meat. These pigs have been domesticated over the centuries from wild pigs.

Wild pigs are strong and self-protective. They live in many parts of the world, mostly in forested areas, and include the warthog from Africa, and the wild boar from parts of Europe, Asia and Africa. (The European wild boar is the ancestor of the domestic pig.) Wild pigs have a coat of coarse hair, which is generally brown or grey in colour.

Most wild pigs make a short burrow, roughly lined with twigs or grass, in which they hide or rest. Sows (female pigs) produce their litter of five or six young in their burrow. The piglets are helpless at birth and

wild boar

warthog

remain in the burrow for the first few days of their life.

Wild pigs are generally peaceable animals, but if they are attacked they defend themselves with their large, sharp canine teeth (tusks). These tusks curve up out of the pig's mouth.

Pigs feed on grasses and small plants, and often use their long snouts, which are strengthened with a special bone, to dig up roots. However, pigs are not vegetarians, and farm pigs are often fed on scraps which can include fish or meat, as well as corns and grains.

• In spite of their reputation, pigs are not exceptionally dirty. However, they may sometimes wallow in mud, probably to keep their skins cool.

• The peccaries of Central and South America look and behave rather like pigs, but they are not closely related. They move about in large groups which may turn as one on an intruder.

find out more
Mammals
Skin

Plants

Plants

◀ The coastal redwood is the tallest living tree and can grow up to 110 metres tall. It belongs to the group of conifers, which are recognized by their cones and needle-like leaves. Its height helps to ensure that these leaves reach sunlight for photosynthesis.

• The study of plants is called botany, and people who study them are called botanists.

Plants range in size from giant trees to tiny mosses. They include the largest and the oldest living things. But unlike animals, plants do not usually move around. Instead, they grow rooted in the soil or attached to the ground or to rocks.

Plants do not feed like animals, although they have to take in chemicals from the soil or from their surroundings to grow. Instead of eating, most make their own food in special 'chemical factories' in their leaves.

Green factories

The most important difference between plants and animals is how they get their food, and this also helps to explain the shape and form of a typical plant. Most plants are green in colour and have leaves. These leaves are spread out in a way that catches most sunshine.

Plants, like animals, are made up of microscopic 'building blocks' called cells. The plants are green because their cells contain a green chemical called *chlorophyll*, and the chlorophyll is bundled into packages called *chloroplasts*. The chloroplasts are the chemical factories that make the plant's food. To do this, the plant also needs two 'raw materials' from its surroundings. These are the gas carbon dioxide, which is common in the air, and water. The carbon dioxide passes into the plant's leaves through special pores in their surface, while the water is taken up by the plant's roots. The

chlorophyll in the plant's leaves absorbs energy from sunlight, and this energy is used to turn carbon dioxide and water into various kinds of energy-rich sugars. These sugars are the plant's food. The whole process of making food using sunlight is called *photosynthesis*.

A waste product of photosynthesis is the gas oxygen. This is fortunate for humans and other animals, because we need oxygen to breathe. In the process, we breathe out carbon dioxide, which would eventually poison us, if plants did not turn it into oxygen. Plants therefore have an important role in recycling the air that we breathe.

Food stores

The sugars produced by plants in photosynthesis can be joined together and stored as more complex chemicals called starches. These are often stored in parts of the plant's roots or underground stems, which swell up into bulging growths called tubers. Sometimes we eat these starch stores as food. Carrots, for example, are swollen roots, and potatoes are swollen underground stems.

The plant can break down the starch and the sugars whenever it needs energy, for example at night when there is no sunlight to provide energy, or when it needs extra energy to grow or to produce seeds. Plants also need other chemicals (called *minerals*) to build all the parts that make up their cells. These chemicals are taken in by the plant's roots, dissolved in water.

▼ (1) Roots reach down into the soil to gather the water and minerals the plant needs for growth. (2) The roots of some plants swell up and are used to store the plant's food, as in a carrot.

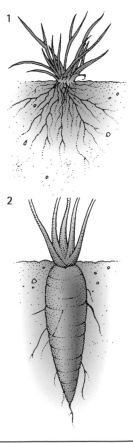

1

2

Plant variety

We can think of a tree as a 'typical' plant. Its roots anchor it in the ground and absorb water and minerals. The main job of its stem or trunk is to hold up the leaves, so they can collect sunshine and carbon dioxide for photosynthesis. Tube-like cells inside the stem carry water from the roots to the leaves, and other tubes transport sugars to wherever the plant needs them for energy. Twigs and branches spread out the leaves so they can best collect sunshine. Plants such as mosses and liverworts are much simpler in design, but they still photosynthesize and they still fit the basic pattern of a plant.

Some plants are not green: they contain no chlorophyll and cannot photosynthesize. These plants get their food in different ways. Plants such as the 'stinking-corpse lily', *Rafflesia*, are parasites – they 'steal' from other plants. Their roots grow into the roots or stem of another plant and draw off the food and water they need to live and grow. Other plants such as toothwort get all the goodness they need from dead plant material in the soil. Their roots release chemicals which break up fragments of dead leaf or twig in the soil to release sugars and minerals, which the plant then absorbs through its roots.

There are some groups of living things which are very plant-like, but which scientists do not regard as true plants. Many algae, for example, are made up of a single cell which contains chlorophyll and can photosynthesize. But some algae move about like animals, and so algae are

▲ Mosses have a simpler structure than flowering plants. They have no proper roots, and collect all the water and minerals they need from the rain that falls on them.

generally regarded as a separate group (or kingdom) of living things (even though seaweeds are giant algae which look very much like plants). Mushrooms and toadstools also look like plants, but they are actually the fruiting body of a fungus, which is mostly made up of a tangle of tiny threads in the soil. The fungi are therefore put in another kingdom. The plant kingdom includes mosses, liverworts, ferns, conifers and true flowering plants.

Spreading the kind

Because plants do not move about like animals, they cannot find their way to new places where it might be better to live, and they cannot meet and mate with others of their kind, as animals do. Plants therefore need outside help both to breed and to spread themselves. Flowering plants produce male and female sex cells in different parts of their flowers. They rely either on the wind or on animals to carry the male sex cells (called pollen) to the female sex cells of another flower. There, the sex cells can join, in a process called fertilization, to produce seeds.

Seeds allow new plants to grow, and help to spread their kind. They are usually enclosed in a fruit. The fruit is designed to carry the seed away from the parent plant. This helps the plant invade new areas, and ensures that the young plant

▼ Flowers like the ones on these cacti are designed to attract insects, which then help the plant by carrying pollen to another flower so that fertilization can occur. The shape and colour of different flowers attract particular insects. This helps to ensure the pollen is carried only to the right kind of plant.

Herbs

We normally think of herbs as plants, like mint and parsley, whose leaves are used as flavourings for food. But in botany, a herbaceous plant is one that does not have a woody stem, and dies right down to the ground after flowering. Not all the plants that we use as herbs are actually herbaceous. For example, bay leaves are herbs that grow on a small tree.

does not compete with its parent for space and sunlight.

Different life cycles

Other plant groups spread and multiply their kind in slightly different ways. Conifers, for example, also have tiny flowers that spread pollen for fertilization. But their seeds are usually produced on a woody cone that does not completely enclose the seed. Some conifers, such as yew trees, have juicy berries instead of cones, but the seed is still not completely enclosed. The seeds from cones are spread by the wind, while the berries are spread by animals.

Ferns, mosses and liverworts have more complicated lifecycles. Instead of seeds they spread by *spores* – dust-like particles that are spread in the wind. The spores grow into young plants that are quite unlike the adults; they are tiny green flaps that can only survive in damp places.

▶ Mistletoe grows attached to branches, high in a tree. Its roots penetrate the branches and collect minerals and water from the tree, but its green leaves still make their own food by photosynthesis. Its sticky white berries are spread to other trees by birds.

These produce sex cells in tiny pockets on their underside. The male sex cells swim through a film of moisture on the surface of the flap to the pockets containing the female sex cells, which they then fertilize. An adult fern, moss or liverwort then develops from the fertilized cells.

find out more
Algae
Flowering plants
Forests
Fungi
Grasses
Seaweeds
Trees and shrubs

▼ Plants are divided into different groups, according to how they reproduce.

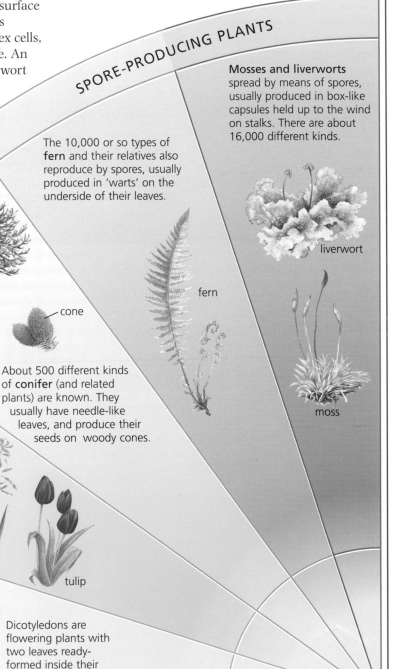

SPORE-PRODUCING PLANTS

Mosses and liverworts spread by means of spores, usually produced in box-like capsules held up to the wind on stalks. There are about 16,000 different kinds.

liverwort

The 10,000 or so types of **fern** and their relatives also reproduce by spores, usually produced in 'warts' on the underside of their leaves.

fern

moss

CONE-BEARERS

monkey-puzzle tree

cone

About 500 different kinds of **conifer** (and related plants) are known. They usually have needle-like leaves, and produce their seeds on woody cones.

FLOWERING PLANTS

monocotyledons

The 240,000 known types of **flowering plant** are divided into two groups, depending on the number of leaves that spring directly from their seeds. Monocotyledons produce a single leaf inside each seed.

grass

tulip

dicotyledons

Dicotyledons are flowering plants with two leaves ready-formed inside their seeds.

oak tree

sunflower

Platypuses *see* Mammals

Pollution

We cause pollution when we do damage to our surroundings. Leaving litter is one form of pollution. But chemicals and waste from factories, farms, motor cars and even houses cause much more serious pollution.

Pollution damages humans and other living things. Many countries are now trying to limit this damage.

Air pollution

Factories, power stations and motor vehicles make waste gases, which are released into the air. The polluted air damages the health of people, plants and animals. Waste gases in the air can also cause acid rain, which damages trees, lake and river life, and buildings.

Water pollution

Acid rain can pollute lakes and rivers. But there are many other kinds of water pollution, including untreated sewage and poisonous chemicals from factories. These pollutants can kill fishes and other water animals and plants. Farmers use fertilizers and pesticides that are washed into rivers and streams after rainfall.

Pollution on land

People cause pollution when they dump their rubbish. Some kinds of rubbish rot away, but many plastics will never decay.

Radioactive waste from nuclear power stations could cause very serious pollution. Nearly all nuclear waste is safely contained for now, but small amounts do sometimes escape into the air or water. Many scientists are worried about the long-term effects of this type of pollution, because nuclear waste can stay radioactive for thousands of years.

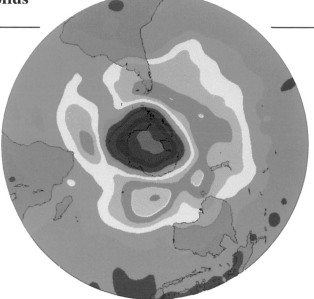

▲ A thin layer of ozone surrounds the world, and protects us from the harmful ultraviolet rays in sunlight. Ozone is destroyed by burning fuels and by chemicals called CFCs released from refrigerators and aerosols. This satellite picture taken above the Antarctic shows ozone levels in the upper atmosphere. In the blue, purple and grey area at the centre there is virtually no ozone at all.

find out more
Conservation
Ecology

Ponds

A pond is a small, shallow body of water. Some are created by farmers and gardeners. Others form naturally in wet hollows and ditches, fed by rain or melting snow. Ponds attract animals to drink, and birds coming to look for food.

Ponds are usually shallow, so the water contains plenty of oxygen for the pond animals to breathe. Tiny oxygen bubbles are also produced by pondweeds as they photosynthesize (make food using sunlight). The surface of the pond heats up by day and cools down quickly at night. In shady, deeper water the temperature changes less.

Most ponds change with the seasons. In winter, the floating plants sink to the bottom, or are torn up by the wind. In cold countries, a winter cover of ice protects the deeper water below from cold air, allowing frogs, toads and fishes to survive at the bottom of the pond.

Disappearing ponds

If left alone, many ponds eventually silt up and disappear. In hot climates the pond may shrink or even disappear during the dry season, as the water evaporates.

In recent years, many village ponds have been drained so that the land can be built on or farmed. Others have been filled in as rubbish dumps or become polluted by fertilizers or pesticides.

find out more
River life

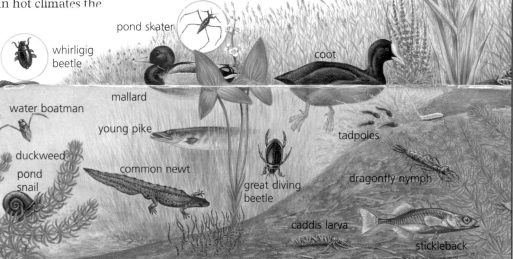

yellow iris
damselfly
pond skater
whirligig beetle
coot
mallard
water boatman
young pike
duckweed
pond snail
common newt
great diving beetle
tadpoles
dragonfly nymph
caddis larva
stickleback

Prehistoric life

- Ninety-nine per cent of all the different kinds of animal that have ever lived are now extinct.

- An extinction is the total disappearance of a particular species (kind) of plant or animal. Although there have been several great extinctions, very few phyla (major groups) of animals have disappeared completely.

▼ Evolution of life began more than 3500 million years ago. This time-line shows only the last 550 million years. Scientists can never be sure of the exact time an animal or plant evolved, so the time-line shows only when something was living rather than when it first appeared. Some of the kinds of animals that lived 500 million years ago, such as the starfishes, are still living today, but most are now extinct. (These drawings are not to scale.)

The term 'prehistoric life' means all the living things that existed before humans began to write about their history. We know about some plants and animals only from their fossils as they lived before human beings. Cave paintings and physical remains tell us about other animals that were well known to our ancestors. Many of these animals are now extinct.

When the Earth was first formed, about 4600 million years ago, there was no life at all, for its surface was too hot for any living things to exist. Scientists think that life first started in the muddy surfaces heated by volcanoes. The oldest fossils known date back about 3500 million years. They are much like bacteria of the present day.

Life in the sea

Hundreds of millions of years passed and some of the living things began to use the Sun's energy to make food. In this process, called photosynthesis, they used carbon dioxide in the atmosphere and gave off oxygen as a waste product. Plants still use this method of making food today. There is evidence that very small and simple plants existed about 3000 million years ago. As a result of plant activity, oxygen gas became part of the Earth's atmosphere, gradually building up to levels similar to those found today.

The first animals were probably like some single-celled creatures of today, such as Euglena, which makes a green scum on ponds. As time

▲ These living stromatolites in Australia are made up of layers of *cyanobacteria* (blue-green algae) and rock. The oldest known fossils are of stromatolites 3500 million years old.

went on, the single cells combined to make larger life forms. The first chains of plant cells date back about 1000 million years, and geologists have found the remains of some of the first complex animals in rocks that date back over 600 million years: first sponges, then jellyfishes and worm-like creatures.

About 570 million years ago many kinds of animal developed hard shells, which became fossilized much more easily than the soft bodies of earlier creatures. Fossils of this age include

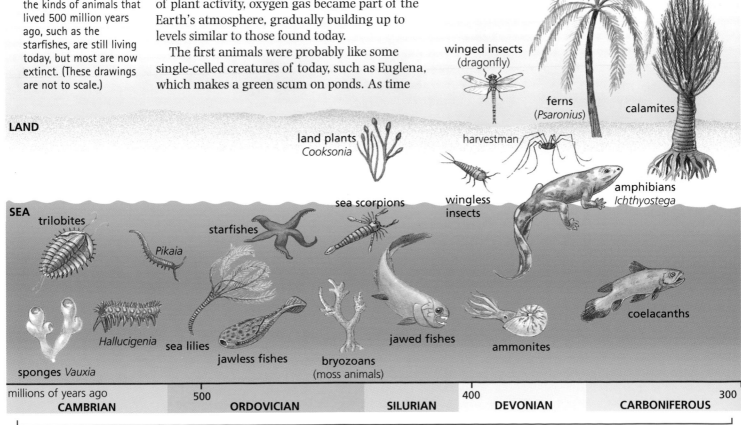

winged insects (dragonfly)

ferns (*Psaronius*)

calamites

harvestman

LAND

land plants *Cooksonia*

wingless insects

amphibians *Ichthyostega*

SEA

trilobites

Pikaia

starfishes

sea scorpions

sponges *Vauxia*

Hallucigenia

sea lilies

jawless fishes

bryozoans (moss animals)

jawed fishes

ammonites

coelacanths

millions of years ago			500						400					300
CAMBRIAN				ORDOVICIAN			SILURIAN			DEVONIAN				CARBONIFEROUS

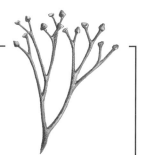

corals, brachiopods, starfishes, snail-like animals, and most important, trilobites, which were armoured creatures related to crabs and insects.

The first fishes, whose remains are found in slightly later rocks, were armoured with heavy bone to protect them against hunting invertebrates (animals without backbones). Eventually many kinds of fish, including sharks and the ancestors of bony fishes, lived in the seas and fresh waters.

Life on land

From the fresh waters, plants and animals began to invade the land. This began about 400 million years ago, when a few small plants, fungi, mites and insect-like creatures lived on the muddy shores of estuaries and lakes. Some of the animals fed on the plants. In time plants grew larger and covered much of the land, making suitable habitats for many kinds of animal.

Unlike most modern fishes, many fishes at this time could breathe dry air, for they had lungs as well as gills. Some had pairs of large fins, supported by leg-like bones. It is thought that these fishes lived on the bed of pools or rivers, using their fins like legs to walk over the mud. They were all flesh-eaters, and it is likely that they sometimes scrambled right out of the water to grab prey such as an insect or snail from the bank.

In time, amphibians developed from such 'walking fishes'. Amphibians (which include today's frogs and salamanders) are able to survive out of water for at least part of their lives, although they have to return to water to breed. The first amphibians dragged themselves over the ground. But a few developed stronger legs, and were able to lift themselves clear of the rough surface. The descendants of these developed into reptiles. ▶

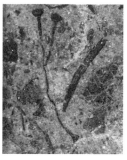

▲ *Cooksonia*, one of the first land plants. The photograph shows a fossil of the plant. The illustration shows how scientists think it looked. It had a waxy outer layer (cuticle) to prevent it drying out, and tubes in the stems to transport water and food.

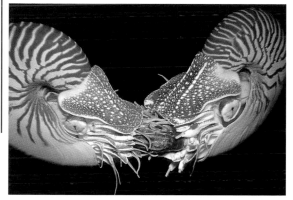

▲ Nautiluses, like the two in this picture, are sometimes called 'living fossils'. Their ancestors, known as nautiloids, include ammonites and were common 450 million years ago, during the Ordovician period.

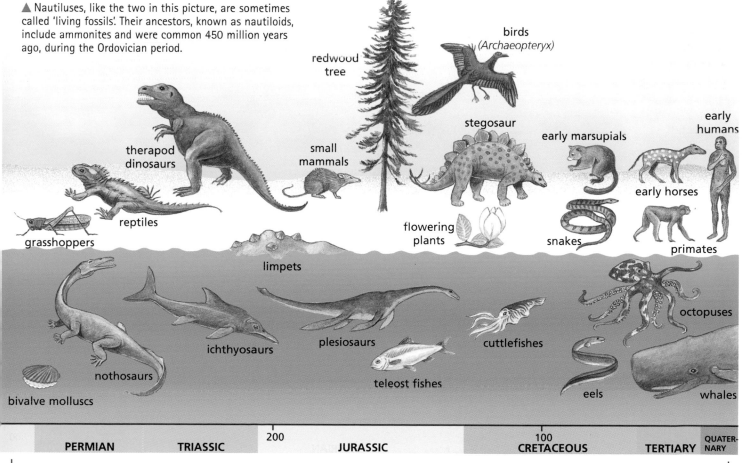

redwood tree

birds (*Archaeopteryx*)

therapod dinosaurs

small mammals

stegosaur

early marsupials

early humans

reptiles

flowering plants

early horses

grasshoppers

snakes

primates

limpets

nothosaurs

ichthyosaurs

plesiosaurs

cuttlefishes

octopuses

bivalve molluscs

teleost fishes

eels

whales

PERMIAN	TRIASSIC	200 JURASSIC	100 CRETACEOUS	TERTIARY	QUATER-NARY

The first mammals and birds

Most reptiles were heavy-bodied and slow-moving, but a few developed in such a way that they were able to move faster and so escape from predators (hunting animals). Changes in their skull and their ribs suggest that they breathed regularly, like mammals. The first true mammals had developed from these mammal-like reptiles by the time of the earliest dinosaurs.

Birds evolved later. The earliest bird that we know of lived about 147 million years ago. It was clearly related to small flesh-eating dinosaurs.

When the dinosaurs finally died out about 65 million years ago, mammals, birds and modern types of fishes and flowering plants survived. In a relatively short period of time huge numbers of mammals and birds occupied much of the land and the seas. In time the mammals were left as the largest living creatures.

The coming of humans

One of the most recent mammals to evolve was humans. Our ancestors in prehistoric times shared their world with many animals that are now extinct. We know about these animals from cave paintings, engravings, fossils and other remains.

At the end of the last ice age in Europe, hunters in France and Spain painted pictures of animals on the walls of caves. Most were of wild horses, deer and bison, which were their main food. Other animals well known to our ancestors but now extinct are the mammoth, the woolly rhinoceros, the cave lion and the cave bear.

Most of what we know about prehistoric animals is from their fossilized remains, but a few creatures have been completely preserved, and we can see their muscles and skin. In parts of Siberia and North America woolly mammoths fell into crevasses in the frozen ground and were unable to get out. These animals were in a natural deep-freeze, in which they were preserved for over 10,000 years.

In North America the remains of a kind of extinct elephant called a mastodon tell us that these creatures were sometimes hunted successfully by humans. One fossil skeleton has been found with a small stone implement stuck into one of the bones.

● All domestic animals, including our pets, cows, horses and pigs, are descended from wild species. People in prehistoric times caught and tamed animals that could be useful to them. Tamed wolves helped in hunting; cattle, horses, sheep, goats and camels were kept for meat, milk, fur and hide and were often used to transport loads.

find out more
Animals
Cells
Dinosaurs
Evolution
Fossils
Plants

◀ After insects, pterosaurs were the first animals to fly. They were reptiles that had gradually developed wings over millions of years, taking to the skies over 200 million years ago. They became extinct 70 million years ago. Birds were the next animals to fly, and finally bats, which are the only flying mammals.

The first reptiles

The first reptiles lived about 280 million years ago. Their skin was probably hard, dry and waterproof, and their eggs probably had a shell like those of modern reptiles.

Slowly the early reptiles spread across the continents, adapting to many different ways of life. Some returned to the water. Plesiosaurs and pliosaurs swam like the seals of today, while ichthyosaurs were the equivalent of dolphins. Great lizards called mosasaurs also lived in the seas. Flying reptiles soared above. Dinosaurs were among the numerous reptiles on land. Some of these were larger than any other land animals, before or since.

▼ These sabre-toothed cats ranged over North and South America about 2 million years ago, and probably fed on large mammals. It is thought that they attacked the soft underbellies of their prey, using their teeth to remove its innards.

Primates

'Primates' is the name given to the group of mammals to which human beings belong. It also includes apes, monkeys and various 'lesser primates', including the lemurs of Madagascar, the pottos of Africa and the lorises of Asia.

Except for humans, most primates live in tropical forests. When they climb trees, they hold on to the branches with fingers and a thumb that bends round to meet the fingers. This is called an 'opposable' thumb. They also have an opposable big toe. This enables most primates to grip things very strongly with both their hands and their feet.

Eyesight is a primate's most important sense. Its eyes are at the front of its face, so it looks at things using both eyes together. This is called binocular vision, and it enables primates to judge distances accurately. This is very important when they leap among the branches. Most primates do not have a good sense of smell, though, as a large nose would get in the way of their vision.

Most primates are social creatures that live in family groups. They have large brains, and are more intelligent than most other animals. In general, primate mothers have only one baby at a time. It is usually cared for by its mother and learns both from her and by playing with other members of the group.

Many primates have a long tail. In most cases it is used only for balancing, though some South American monkeys have a prehensile (gripping) tail that they can use like an extra hand. Humans, like their closest relatives the apes, have no tail.

▲ A slender loris from southern India, so called because it has long, slender arms and legs. It is a nocturnal animal (active at night), spending its nights using its huge eyes to look for insects and other small animals. It spends the days curled up asleep in the trees.

• Lemurs, lorises and pottos are known as prosimians, or lesser primates. Prosimians have several features that distinguish them from other primates, including a wet nose with slit-like nostrils.

find out more
Apes
Human beings
Mammals
Monkeys
Prehistoric life

Proteins

▼How proteins are made.

Proteins are among the most important substances that make up living things. They can carry out almost any task that a living cell requires.

1 The body's genes carry the information for making many proteins on a long molecule called DNA.

DNA

mRNA

2 One gene, with the instructions for making an individual protein, is copied to another molecule, called mRNA (messenger RNA).

3 Tiny particles called ribosomes use the instructions from mRNA to join a string of amino acids together in a specific order.

ribosome

4 Once a protein chain has been formed, it folds itself into a particular shape.

protein chain of an enzyme

5 Some more complex proteins are made up of several protein chains.

chain A

chain B

chain B

chain A

find out more
Cells
Digestive systems
Genetics
Glands
Viruses

Proteins are large complicated molecules. They are built of smaller units called amino acids, strung together in chains. An average protein is about 500 amino acids long. All the proteins found in living things are made from only 20 different amino acids.

There are two main types of protein. *Structural proteins* make up the bulk of such body tissues as muscles and ligaments. *Enzymes* are proteins that help chemical reactions in the body to occur quickly and efficiently. In digestion, for example, enzymes help break down food into smaller parts so that it can be absorbed into the body.

The different functions of structural proteins and enzymes depend on their shapes. The amino-acid chains of structural proteins make long, straight shapes that can join together in threads, bundles or sheets. But in enzymes, the chains fold up into complex shapes, each one designed to help a particular chemical reaction.

Rabbits and hares

Rabbits and hares are small furry mammals with long ears. In general, hares have bigger ears and larger bodies than rabbits, who are their close relatives.

• There are 21 different kinds of rabbit and 22 kinds of hare.

find out more
Mammals

Rabbits and hares are so similar that in some parts of the world creatures that should be called hares are known as rabbits. For example, some kinds of North American hare are often called rabbits or jackrabbits.

Rabbits

Rabbits are gnawing animals, feeding on plants. They usually live in groups and shelter in burrows. When they feed, they stay close to home for safety. They rely on their good eyesight and senses of hearing and smell to warn them of enemies. They are hunted by many predators, including human beings.

Baby rabbits are blind and furless at birth. Although their

◀ During the mating season male hares leap, chase and 'box' each other to attract the attention of females.

mothers feed them only twice a day, they grow quickly. They are ready to face the outside world by the age of 3 weeks. Most kinds of rabbit produce a large number of young in a year.

Hares

Unlike rabbits, hares live alone, usually in grassy places. They do not dig burrows, but make a shallow trench, called a form, to shelter in. A hare lies quite still when a dog or a fox approaches. It leaps away only when the enemy is very near. It can run very fast, zigzagging and jumping high, so that it can still watch its enemy.

Baby hares are called leverets. They are open-eyed, furry and active within minutes of being born. Their mother hides them near to her form and feeds them every evening, but they are soon able to look after themselves.

Racoons

Racoons are among the most familiar animals in the Americas. Many racoons have discovered that humans produce lots of edible waste. They often make their dens in lofts and outhouses to be near to dustbins and rubbish tips.

• The name 'racoon' comes from 'aroughcoune', a Native American name for the animal. It means 'he scratches with his hands'.

• Racoons are best known for their striped tails, which were worn by hunters of the American West, such as Davy Crockett.

• There are seven members of the racoon family.

find out more
Mammals

Most racoons live in forests, near water, and they climb and swim well. They are more active in the evening than in the day. They eat almost anything: wild fruit, birds' eggs and insects in the forests, worms, fish, frogs, young muskrats, crayfish and clams in the water. They have front paws like little hands, with long, mobile fingers that can

break open shells and even undo catches and locks.

Normally racoons live alone, and a male may travel a long distance to find a mate. Most females have a family of four or five young, born in April or May. The young develop slowly. Born blind and helpless, they do not leave the den until they are about 8 weeks old, and are not weaned until the end of the summer.

The red panda is the only member of the racoon family to live outside the Americas. It lives in mountain forests in China and the Himalayas. Red pandas are good climbers. They sleep by day in tree dens and are active at night, searching for fruits, acorns and plant shoots, and sometimes catching insects and other small animals.

◀ A racoon raiding a henhouse for eggs.

Reptiles

Crocodiles, tortoises, snakes and lizards are all reptiles. They are cold-blooded, and usually have a tough, dry, scaly skin.

Reptiles are backboned animals (vertebrates) and breathe through lungs. Their body temperature, like that of all cold-blooded animals, goes up and down depending on the temperature of their surroundings. Most reptiles live in the warmer parts of the world. Although they have strong limbs, most reptiles are slow-moving animals.

In the breeding season, males use dances or bright colours to attract females. After mating, females lay their eggs in a sheltered, dry place. Most reptile eggs have a papery or leathery shell, but are otherwise very much like birds' eggs. Most reptiles leave their eggs to hatch in the warmth of the Sun, but some stay near the nest, and others, such as crocodiles and some snakes, look after their eggs and young. A few reptiles keep their eggs inside their bodies until they are ready to hatch. Baby reptiles look just like their parents.

In the past, many kinds of reptile flourished. The dinosaurs were reptiles and included the largest known land-living creatures. Today there are only four groups of reptiles.

◀ *Turtles and tortoises* are all armoured with a bony shell. Many swim well and live in water, but come onto land to lay their eggs.

▲ The *tuatara* lives on a few islands off the coast of New Zealand. It is the only survivor of a big group of reptiles which flourished at the time of the dinosaurs.

▲ *Lizards and snakes* are the best-known and the most abundant reptiles.

◀ *Crocodiles and alligators* are the biggest reptiles alive today. All live in or near water, although they have to come ashore to lay their eggs.

- There are about 6000 different kinds of reptile.

- The largest reptile is the salt-water crocodile, which may grow up to 7 m.

- Big crocodiles and turtles may live to be 200 years old.

find out more
Crocodiles and
   alligators
Dinosaurs
Lizards
Snakes
Turtles and tortoises

Rhinoceroses

Rhinoceroses, or rhinos, are large horned mammals that live in the tropical grasslands and forests of Africa and Asia. All five surviving kinds of rhinoceros are in danger of becoming extinct.

Rhinoceroses are different from other horned mammals in that their horns are not on the top of their heads but towards the end of their noses. And these horns are not bony, like the horns of cattle or the antlers of deer, but are made of compressed thick hairs. Rhinos are related to horses, and not to elephants and cattle.

Rhinos feed entirely on plants. Most rhinos are browsers – they eat leaves and twigs from bushes and trees which they pick using a grasping upper lip. But white rhinos are grazers – they feed on grass, which they crop with their distinctive broad lower lip.

All rhinos have good senses of hearing and smell, but they are short-sighted. They usually live alone and are suspicious of strangers, and may attack intruders without real cause.

Baby rhinos may be hunted by the big cats, but humans are the only enemies of the adults. In the past, dozens of kinds of rhinoceros lived in many parts of the world. Today, only five different kinds survive, and all of these have been hunted and poached to the point of extinction.

find out more
Africa
Grasslands
Mammals

◀ A white rhino and calf. White rhinos are in fact grey brown in colour — the 'white' in their name comes from an Afrikaans word meaning 'wide', and refers to their broad lips.

River life

A river travels from its source in the mountains or other highland area to the estuary where it flows into the sea. In the early stages of a river's journey, many small streams join up to form the main river, with other streams feeding in later on. Rivers and streams provide many different habitats for plants and animals.

Some animals live on the river-bed, others in the open water, among the plants near or on the river bank, or on the bank itself. Herons, kingfishers and otters hunt for fish in the river, and swallows and bats swoop over the water to catch mayflies. The Amazon River alone contains over 2000 different kinds of fish, including piranhas, catfish, electric eels and the giant arapaima. Some river animals such as the Nile crocodile grow very big.

Dead plant and animal material washes into rivers from the land, and leaves and seeds blow into the water. This rotting material provides food for microscopic plants, worms, water snails, shrimps and mussels. These creatures in turn are food for fish, crayfish and turtles, and these animals may be eaten by otters, crocodiles and alligators. In very muddy water, it can be difficult to search for food. Electric catfish and electric eels use tiny pulses of electricity to detect their prey in the water.

River animals must avoid being washed away. Freshwater limpets cling fast to the rocks. Water beetles and mayfly nymphs have flattened bodies for hiding in crevices, and lampreys use sucker-like mouths to cling to rocks.

◀ The Yangtze river dolphin, like other river dolphins, is virtually blind. River dolphins live in muddy rivers in South America, Southern Asia and China. They find their way and detect prey by listening to the echoes of their own clicking noises. The Yangtze river dolphin is the most endangered of all marine mammals. There may be only about 100 individuals left.

find out more
Fishes
Ponds
Wetlands

Salmon and trout

Salmon and trout are closely related fishes that live in seas, rivers and lakes in many parts of the world. They are very popular with anglers, and are also an important source of food. In many places they are reared in fish farms.

Two of the best-known kinds of trout are the European trout and the rainbow trout of North America. There are two kinds of European trout. The first is the sea trout, which migrates from a river to the sea, and then returns to the river to breed. The second is the smaller brown trout, which remains in fresh water all its life. The rainbow trout also has a form that visits the sea, and another that stays in fresh water.

Salmon are generally larger than trout, and most kinds migrate. The sockeye salmon of the northern Pacific Ocean live for some years at sea. Then they return to the waters where they were hatched – their breeding ground – swimming up river, sometimes over 2000 kilometres, to get there. They probably recognize the place by smell. After the female has laid her eggs and the male has fertilized them, both die. The Atlantic salmon has a similar life history, except that it can breed more than once.

• The largest Atlantic salmon ever caught weighed nearly 39 kg. It was landed in Norway in 1928.

◀ An Atlantic salmon leaping up a waterfall on its way to its breeding ground. It is a powerful fish, and can grow up to 1.5 m in length.

find out more
Fishes

Rodents *see* Mice, etc. • **Salamanders** *see* Newts, etc. • **Savannah** *see* Grasslands

Sea birds

Almost three-quarters of the Earth's surface is covered by sea. The oceans are as rich in food as the land, so it is not surprising that birds take advantage of this food supply.

However, the open sea can be a harsh place for a bird and birds must come to land to breed. Only 300 or so of the 8700 different kinds of bird have come to depend on the sea.

Gulls are a particularly successful group of sea birds and many have benefited from humans. They eat our rubbish and some nest on our buildings. Gulls may be seen many kilometres from the coast and some kinds live entirely inland.

gull

Albatrosses and shearwaters are great ocean travellers. The curious shape of their bill gives this group its name, the 'tubenoses'. The tube is an enlargement of the bird's nostrils and it may be used to gauge the strength of the air flow or to help the bird smell land or food. Albatrosses are spectacular gliders and only come ashore to breed.

albatross

• A manx shearwater ringed in Wales was found in Brazil 16 days later. Its minimum flying speed must have been 740 km per day.

◀ Skuas are bird pirates. They are related to gulls and normally feed by following other sea birds and then chasing and harrying them until they drop a fish.

Another group of sea birds includes the cormorants, gannets, boobies and frigate birds. Their methods of finding food vary from the dramatic plunge of the gannet head-first into the water from a height of 15 metres or more, to the 'piracy' of frigate birds, which chase and rob other sea birds. ▶

cormorant

◀ Terns, sometimes called 'sea swallows', are close relatives of the gulls. Some of them are great travellers, like the Arctic tern which migrates from the Arctic to the Antarctic and back each year. Terns feed mainly on fish.

emperor penguin

little auk

Penguins

Penguins are flightless sea birds that live south of the Equator, especially in the Antarctic waters. Penguins have a layer of blubber (fat) below their skin as a protection from the cold. Their streamlined bodies and flipper-like wings help them to shoot along in the water like small torpedoes. Penguins are heavy and this helps them to dive. Emperor penguins can stay under water for up to 10 minutes. Most penguins hunt fish, although some kinds eat squid and krill. Penguins are social animals and live in large colonies.

find out more
Birds
Migration
Oceans and seas

▶ Atlantic puffins close to their burrows. Puffins nest in colonies. They breed from late May. The female usually lays a single egg in an abandoned burrow, or in a hole she digs herself with her feet.

Auks

Members of the auk family are all found north of the Equator. They look rather like penguins, but auks are able to fly. The puffin is a member of this family. Like penguins, auks swim and dive, and feed on fish. Many auks nest in huge colonies on sea-cliffs. Before they are fully grown, the young will flutter from their cliffs and disappear out to sea for the winter. The flightless great auk used to be a common member of this family, but was hunted for its flesh and became extinct in 1844, when the last surviving birds were killed.

Seals

• There are 34 different kinds of seal, sea lion and walrus. The largest is the elephant seal, which sometimes grows up to 6 m long, and weighs up to 5000 kg.

find out more
Mammals
Oceans and seas

Seals and their relatives the sea lions and walruses are sea mammals. Although they spend most of their lives in the water, they breathe air. Most of these animals live in the cold waters of the northern and southern oceans.

▼ Walruses use their huge tusks (up to 1 m long in males) to dislodge clams and other shellfish on the sea-bed. They then gather them with their mobile, whiskery lips.

Seals are slow and clumsy on land, but once in the water they are swift and graceful. Their bodies are streamlined, and they propel themselves through the water using their powerful flippers.

Seals can stay in the water for long periods. Beneath their skin they have a thick coat of special fat called blubber, which helps to keep them warm. When they dive, their heart rate slows to 4–15 beats per minute. This reduces the amount of oxygen that they use.

Most seals hunt fish, often those not valued by humans. A few kinds feed on krill (shrimp-like creatures), and walruses eat shellfish and sea urchins. Leopard seals feed on penguins and other seals.

Seals come to land to produce their young. They haul themselves ashore on islands or isolated beaches, which are traditional breeding places. Females produce only one baby, which grows quickly and is sometimes independent before it is three weeks old.

▼ Grey seals use their long whiskers to detect changes in water pressure as something swims past, so they can catch prey even in murky water.

Seashore

The seashore is where the oceans and seas meet the land. It is the part of the coast from the low-tide mark to just above the reach of the sea. In some parts of the world the tide does not go out very far and the seashore is only a few metres wide. In other places the tides go out up to 3 kilometres, and the seashore is very wide.

▼ Animals and plants on a rocky shore. The lower shore is wet and animals such as sea anemones, starfishes and sea urchins (1) live among the wet seaweeds in the rock pools. These seaweeds include sea kelp (2), serrated wrack (3) and thong weed (4). The rock pools are also home to small fishes such as blennies. Further up the shore seaweeds such as bladder wrack (5), knotted wrack (6) and spiral wrack (7) cling to the rocks. Mussels (8) fix themselves to the rock with tough threads. Crabs, sea slaters and shrimps survive by hiding in crevices. Limpets, periwinkles and barnacles (9) on the upper and middle shore have strong shells to protect them from the waves and the drying effect of the wind, and strong muscles to grip the rock. Channelled wrack (10) and lichens (11) flourish on the drier rocks.

The lower shore is covered by sea water for much of the day, while the upper shore is covered for only a short time. The upper shore mostly dries out between tides. The animals and plants living there must be able to adapt to both wet and dry conditions, and to put up with rapid changes in temperature.

Rocky shores

Rocky shores have many pools, each with its own community of animals. The rocks are often covered with seaweeds. Seaweeds also fringe the rock pools, providing shade from the sun, and a place where small animals can hide from predators such as seagulls.

Most of the animals on rocky shores breathe oxygen from the water using gills, so they come out to feed when the tide is in. Crabs scavenge for seaweed and animal remains. Barnacles and shrimps filter food from the sea water using bristles on their legs. Mussels sieve the water with their gills, and limpets and periwinkles graze on the seaweeds.

Sandy and muddy shores

A sandy shore often appears to have very little life, but below the surface live a variety of burrowing animals. On muddier shores, there may be as many as 100,000 animals per square metre in the mud, out of sight of predators and protected from the sun and wind. The lugworm (sandworm) lives in a U-shaped burrow. It takes in water and food at one end, and pushes out waste at the other. Burrowing cockles and razor-shells sieve their food from sea water. Tube worms make themselves tubes of sand and mud, and put out sticky tentacles to trap food from the water. Small shrimps and crabs venture out to feed at high tide, but burrow in the mud before the tide goes out, so they are not swept away.

Where rivers meet the sea, there are huge expanses of mud. These shores can support an incredible amount of life. In just a square metre of mud, up to 1000 ragworms, 42,000 spire-shell snails, or 63,000 mud-burrowing sand-hoppers have been recorded (although not all together). Muddy shores and estuaries also attract large numbers of seagulls and wading birds (shore birds), which come to feed on the seashore animals.

find out more
Crabs and crustaceans
Fishes
Jellyfishes and corals
Mussels and oysters
Oceans and seas
Sea birds
Seaweeds
Starfishes and sea
 urchins
Wading birds

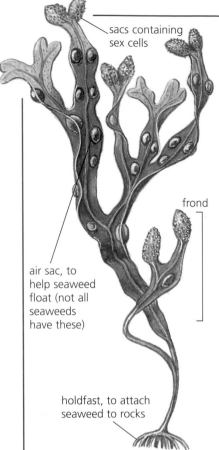

sacs containing sex cells

frond

air sac, to help seaweed float (not all seaweeds have these)

holdfast, to attach seaweed to rocks

◀ The structure of a typical seaweed.

Seaweeds

Seaweeds are the plants you see growing on seashore rocks or washed up onto beaches after storms. They are not truly plants: they belong to a group of living things called algae.

The tiny cells that make up a seaweed are very simple. Their walls are very thin and do not provide much support, so seaweeds can only stand upright when supported by water. Like plants, seaweeds make their food from water and carbon dioxide gas from the air, using the energy of sunlight. This process is called photosynthesis. So seaweeds can only grow near the water surface, where there is enough sunlight for photosynthesis.

Reproduction

Some seaweeds spread by breaking off fragments which grow into new plants. However, seaweeds can also reproduce sexually. The male and female sex cells are found in bladder-like sacs near the tip of a frond. The fertile sex cells are squeezed out through pores and mix together to form a fertilized egg cell. This grows into a new plant.

Zonation

Seaweeds living on shores between the tides have to cope with periods out of the water. Large kelp seaweeds growing at the bottom of the shore can survive out of the water for only a short time. Seaweeds at the top of the shore can survive drying for long periods. Between these two extremes, different seaweeds grow in bands or zones down the shore.

Useful seaweeds

Seaweeds gathered from the shore are used as fertilizers by farmers in many coastal areas. Glue-like substances called alginates, made from seaweed, are used in the food industry to help substances like ice-cream to 'set'. The Chinese and Japanese cook with seaweed. It is considered a very healthy food.

• Most seaweeds are brown or reddish in colour, although a few are green.

find out more
Algae
Plants
Seashore

Senses

Your senses tell you what is going on in the world around you. Human beings, like all animals, need this information to survive – to move and communicate, to find food and escape from danger.

• Different animals have developed senses that equip them to survive well in their surroundings. Bats have such good hearing that they can fly in the dark and catch insects by sensing the echoes of bursts of extremely high-pitched sounds.

Most animals have five main senses: sight, hearing, touch, smell and taste. Each of these senses is located in a part of the body, a sense organ. These sense organs pass information along nerves to the brain. The brain collects all the messages together to produce its 'picture' of the outside world.

Animals see with their eyes. Sight tells them what things look like and where they are. Some animals, including humans, can see colours.

Animals hear with their ears. Hearing helps many animals to communicate, and tells them where a sound is coming from. In humans the main organ of touch is the skin, but in some other animals whiskers or antennae (feelers) are important. Touch tells an animal what things feel like.

Many animals use their nose and tongue to smell and taste, but insects use their antennae for smelling, and organs on their legs for tasting. Humans use smell and taste to help them choose food. Most other animals have a much better sense of smell, and also use smell to communicate with each other.

▶ This evening primrose flower has been photographed in ultraviolet light to reveal the 'honey guides', the dark lines and patches that guide bees to the nectar inside. Bees are able to see these lines, which are usually invisible to the human eye.

find out more
Bats
Brains
Ears
Eyes
Noses
Skin

Sex and reproduction

The populations of many living things are divided into two sexes, male and female, and they reproduce (create new life) sexually. In animals, a sperm cell from the male fertilizes (joins with) an egg cell (ovum) from the female. A new individual grows from this fertilized egg. In most plants the ovule (unfertilized seed) in the female part of one flower is fertilized by pollen from the male part of another flower. This produces seeds that grow into new plants.

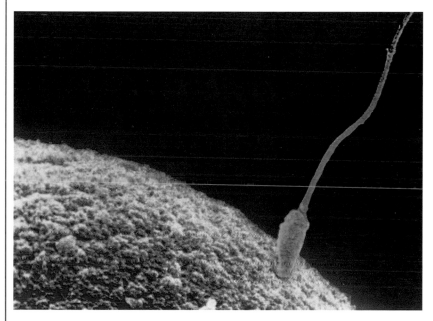

▲ Of the millions of sperm produced during mating, only one will actually penetrate and fertilize each of the female's eggs. Although animals like starfishes may produce millions of eggs, mammals generally only produce a few at a time. Some, such as monkeys, apes and humans, usually only produce one.

Some living things do not have male or female parts, and reproduce non-sexually. This is called *asexual reproduction*. Other living things can reproduce both sexually and asexually.

Asexual reproduction

In asexual reproduction the offspring are identical to their parent and to each other. Most microscopic organisms reproduce asexually. Bacteria and amoebas, for example, simply split in two. Larger, more complicated animals can also reproduce asexually. Some sea anemones and their relatives can reproduce by *budding*. A small bud develops on the side of an adult and grows into a new animal joined to the parent by a stalk. Then the stalk breaks and there are two animals. Many plants can send out side shoots from which new individuals grow, as well as reproducing sexually.

Some kinds of animal can create new individuals from unfertilized eggs. This type of reproduction is called *parthenogenesis*, and in this

way females can have female offspring without the involvement of a male. Several generations of female aphids can reproduce like this. In honeybees a fertilized egg will grow into a male, while an unfertilized egg will grow into a female. Some lizards are also able to reproduce in this way.

Hermaphrodites

Some plants, invertebrates (animals without backbones) and fishes are *hermaphrodites*. This means that they have both male and female parts within them. Most hermaphrodites reproduce sexually with others of their kind rather than fertilizing themselves. Slugs and earthworms, for example, simply exchange sperm when they mate, and both animals then lay fertilized eggs.

Genes and chromosomes

Every cell of every living thing contains genetic material, which is made up of *chromosomes* (coiled strands). Each chromosome contains a large number of genes, and each gene determines how one part of the body is built. In most cells the chromosomes are in pairs. The number of pairs in each cell varies among different plants and animals. In humans there are 23.

But in the sex cells (sperms and eggs) of a plant or animal there is only one set of chromosomes. When the sperm and egg join together to make a new individual, one set of chromosomes comes from the mother's egg cell, and one set from the father's sperm cell.

Male or female offspring?

One of the chromosomes in the sex cell of one parent helps to decide the sex of the young. In human beings, and most other animals, there are two sex chromosomes, called X and Y. All the female's eggs (ova) have an X chromosome, but half the male's sperm have an X and half have a Y chromosome. If a sperm with an X chromosome fertilizes an egg the young will be female. A male young will develop if a sperm with a Y chromosome fertilizes the egg.

The advantages of sexual reproduction

In asexual reproduction, the offspring are all identical to their parent because they have exactly the same genes. In sexual reproduction the offspring, although similar to their parents, are not identical to them or to each other. The offspring receive some genes from the mother, and some from the father. ▶

▲ Some male birds excel at courtship display. Male birds of paradise hang upside-down, showing off their bright plumage to the female bird.

Because of this mixing of genes, sexual reproduction leads to greater variety in a population. This in turn means that a species (type of plant or animal) can adapt itself more quickly to changes in its surroundings. This is because there are always likely to be some individuals that are more suited to the changes than others, and these individuals will survive and reproduce themselves.

Courtship and mating

The physical act between male and female animals that brings about reproduction is called mating. Most animals only mate at certain times of year and they behave in different ways when they are ready to mate. To attract females, male animals often compete with each other in some way. This behaviour is called 'courtship'. Many male birds, for example, perform amazing displays to attract a female. In other animals, such as red deer and elephant seals, the males fight each other to decide which one is the strongest. The strongest male will then mate with all the females in the herd.

An animal's method of mating depends on whether the egg or eggs are fertilized outside the female's body (*external fertilization*) or inside the female's body (*internal fertilization*). Animals that fertilize

externally, such as frogs and fishes, usually get close together before eggs and sperm are released. This form of fertilization is common in water-living animals, but because eggs and sperm do not survive long in dry conditions, most land animals fertilize internally. In internal fertilization the male squirts sperm into an opening in the female's body. The sperm then travel inside her body to reach the eggs. Insects, reptiles, birds and mammals use internal fertilization.

Reproduction in mammals

As with most other animals, male and female mammals (including humans) have different sexual organs (*genitals*). Males have a *penis*. Below and behind the penis is a bag called the *scrotum*, containing two *testes* (or *testicles*). In an adult, these produce several million sperm a day.

The opening in a female mammal's body leading to her reproductive organs is called the *vulva*. A channel called the *vagina* goes from the vulva to the womb (or *uterus*). Two tubes link the womb to the egg-producing *ovaries*. When an egg is produced, it travels down one of these tubes towards the womb, ready to be fertilized. This happens at regular intervals. In humans, for example, the ovaries release an egg about every 28 days.

When mating takes place, the male's penis grows bigger and harder, so that it can be pushed into the female's vagina. The vagina produces fluid to make this easier. Sliding the penis in and out of the vagina leads to *ejaculation*. This is when *semen* (the fluid containing sperm) squirts out from the penis into the vagina. The sperm then swim up towards the ovaries. If a sperm fertilizes an egg, the female becomes pregnant. The fertilized egg lodges in the womb, where it grows and develops until it is ready to be born.

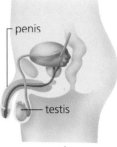

male

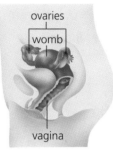

female

▲ The male and female sexual organs of humans are similar to those of other mammals.

• Some living things, including corals, jellyfishes, and plants such as ferns, have a two-stage lifecycle. They reproduce sexually in one generation and asexually in the next. This is called *alternation of generations*.

◀ Like most land animals, the male bear fertilizes the female bear internally, releasing sperm inside her body.

Sharks and rays

Sharks, together with rays and skates, are different from other fishes. The most important difference is that their skeletons are made not of bone but of cartilage – the same material that is found in your nose and ears.

There are many different types of shark. Some, such as the bullhead sharks, are sluggish and feed on shellfish on the sea-bed. Others, such as the massive whale shark, filter the sea water for plankton. The best-known sharks are fast-swimming predators such as the great white. They feed mainly on fish, and the larger ones also eat seals and dolphins.

Sharks' eyesight and hearing are not very good, but they have a keen sense of smell, which they use to find their prey. They can sense vibrations produced by an injured or struggling fish. They can also detect electric currents in the water, which means they can trace prey by the electricity made in its body.

A flesh-eating shark may have up to 3000 teeth, but most of them are behind the front teeth. When one of these is damaged, it drops out and the next tooth moves forward to replace it.

Rays and skates are flat-bodied creatures that live close to the sea-bed. They move along by flapping their large 'wings'. They have a long thin tail, and eyes on the top of their head. Their wide mouths are on the underside of their heads. Stingrays have poisonous spines on their tails, which they use for defence. Other rays, called electric rays, produce an electric shock which they use to stun their prey.

stingray

blue shark

• Most sharks lay only a few eggs, and they are usually protected by horny egg cases. The 'mermaid's purse', which can be found on the seashore, is the empty egg case of a small shark or a ray.

find out more
Fishes
Oceans and seas

◀ The skin of both sharks and rays is covered with thousands of tooth-like structures called denticles. These point backwards, so if you rub the skin from the tail to the head, it feels like sandpaper. Large denticles provide a very strong armour.

Sheep and goats

Sheep and goats are naturally agile, sturdily built animals, related to cattle. They are now rare in the wild, but many domestic varieties have been bred from the wild species.

Most wild sheep and goats live in mountainous areas. Sheep are good climbers but less agile than goats, so they are generally found at lower and less rocky levels. Sheep and goats both have horns, which are larger in the males. They are used particularly in fighting for mates.

In the wintertime wild sheep grow a thick undercoat (fleece) of fine wool to keep them warm and dry. It is moulted (shed) completely in the summer. People started keeping sheep so that this fleece could be shorn off each year and used to make wool. There are now more than 800 breeds and over 800 million domestic sheep.

Wild goats live in herds. They have coats made of thick, often coarse hair. The coats of angora and Kashmir (cashmere) goats are used to produce wool for clothes. Many varieties of goats have been domesticated and kept for their milk, meat and skins.

▶ The alpine ibex is a goat that lives in high altitudes in mountainous regions of Europe. It is a very good climber and can easily leap from rock to rock.

▼ The mouflon from southern Europe and south-west Asia is the smallest of the wild sheep. It was domesticated about 9000 years ago and is the ancestor of domestic sheep.

find out more
Digestive systems
Mammals

• Sheep and goats are ruminants, like cattle and antelopes. This means they eat plants and then regurgitate the food to chew it again.

Skeletons

Your skeleton is the frame of bones that supports your body. It gives you your shape and enables you to make distinct movements. Muscles are attached to the skeleton, and the skeleton and muscles together support the soft parts of the body such as the large organs.

Without skeletons most animals would be floppy objects with no distinct shape, and actions such as swimming, running, picking things up and chewing would be impossible. Skeletons provide animals with a wide range of body shapes, with distinct parts to carry out different functions.

▲ This crab has just shed its cuticle. Cuticles are hard and do not allow an animal to grow. Underneath is a new, soft cuticle. Crabs have to shed their cuticles a number of times before they are fully grown.

Fishes, amphibians, reptiles, birds and mammals (the vertebrates) have internal skeletons made of bone and a softer, gristly substance called *cartilage*. All these animals have a backbone made up of a chain of vertebrae, and a skull. Most also have four limbs – legs, arms, wings or fins.

The skull and backbone support and protect the central parts of the nervous system, the brain and the spinal cord. Attached to the backbone are the ribs; these curve round to form the rib-cage, which protects the heart, guts and lungs. The shoulder and hip are bony connections between the backbone and the limbs. In different land animals the basic five-toed end of the limb has been adapted in many ways. In the horse, for instance, each leg has only a single functional toe, with a hoof (the toe-nail) at its tip.

Other skeletons

Animals without backbones (invertebrates) do not have bony skeletons. Worms and snails, for instance, have taut, fluid-filled bags inside their bodies which help to support them.

Other invertebrates such as crabs, centipedes, lobsters, spiders and insects have a skeleton on the outside of their body. Their soft bodies are supported by a stiff, jointed skin (*cuticle*), which gives them their shape.

Joints

Joints are the bending and sliding places where bones meet. The bones at joints are covered in smooth cartilage, lubricated by joint fluid, which allows them to move over each other easily. Different types of joint allow different types of movement. *Hinge joints*, such as the elbow and the knee, allow movements in two directions only. Where greater flexibility is needed, as in the shoulder and hip, *ball-and-socket joints* are found.

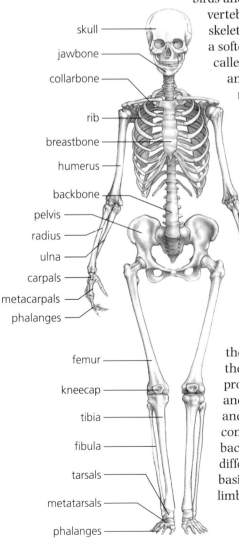

- skull
- jawbone
- collarbone
- rib
- breastbone
- humerus
- backbone
- pelvis
- radius
- ulna
- carpals
- metacarpals
- phalanges
- femur
- kneecap
- tibia
- fibula
- tarsals
- metatarsals
- phalanges

◄ The human skeleton.

find out more
Animals
Bones
Muscles
Teeth

Skin

Your skin is the protective outer surface which covers the whole of your body. Skin prevents the body from losing water, protects it against infection and is an important organ of touch. The bodies of most animals are covered by a layer of skin.

The outer layer of the skin, called the *epidermis*, is made up of dead skin cells. Its cells are constantly being worn away and replaced from the layer of living cells below called the *dermis*. This layer contains blood vessels, nerves, hair roots, and sweat and oil glands.

Your skin helps to control your body temperature. It produces sweat to cool the body down and makes hair to keep you warm. There are also many nerve endings in the skin which send information about touch to the brain.

The thickened skin of some animals, such as rhinoceroses and crocodiles, provides them with a kind of armour. The protective 'shells' of crabs and lobsters are actually a very stiff kind of skin. They also work as external skeletons, and give the animal's body a rigid shape.

• Leather is a material made from the skins (hides) of animals. Most leather is made from domestic animals such as cattle, sheep, goats, pigs, horses and camels. Leather is used to make many things, such as shoes, clothes and furniture.

find out more
Hair
Human beings

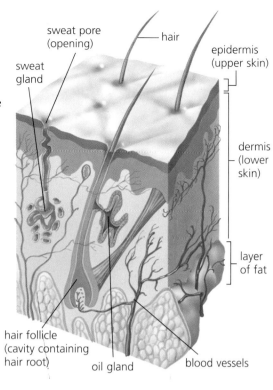

▲ A small piece of human skin shown many times its actual size. The upper dead layer of skin wears away every time you touch something.

Labels: sweat pore (opening); hair; sweat gland; epidermis (upper skin); dermis (lower skin); layer of fat; hair follicle (cavity containing hair root); oil gland; blood vessels

Slugs and snails

Slugs and snails form a highly successful group of animals that are found in large numbers both on land and in water. As a group, they are known as gastropods, which means 'stomach-footed' animals. This is because they all have a large muscular 'foot' which they use to walk or swim.

The main difference between snails and slugs is that in times of danger a snail can pull its soft body into a shell, which is usually coiled. Slugs may have a small shell, or an internal shell, but they cannot pull themselves into it for protection.

In most other ways, snails and slugs are very similar. Land slugs and snails ease their way on their foot with slime, which can be seen after the animal has passed. Both snails and slugs have simple eyes, which are sometimes on tentacles.

Both slugs and snails have a mouth on the underside of the head. This contains very many tiny teeth. Garden snails and slugs eat mostly decaying plants. You rarely notice the damage they do in the countryside, or in an untidy garden, where there is plenty of waste. It is only when there is no dead and dying material that they eat tender seedlings and young leaves.

Snails and slugs live in the sea as well as on land. Sea slugs are usually very brightly coloured, to warn other animals that their flesh is unpleasant-tasting and that they are best left alone. Unlike their land cousins, sea slugs do not feed on plants but on other animals.

• There are more than 35,000 kinds of slugs and snails.

▼ A garden snail may have up to 14,000 teeth in its mouth. They are arranged in rows on a ribbon-like tongue, which is called a radula. This works like a file, rasping away at food.

▼ Sea slugs are almost all brightly coloured. Many of them feed on jellyfishes and use the powerful stinging cells of their prey to protect themselves.

▼ Whelks are snails that live in the sea. They feed on small fishes and other sea creatures, which they suck up through their long proboscis (snout).

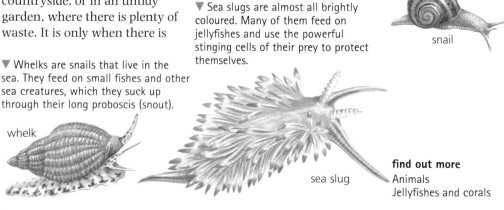

Labels: snail; whelk; sea slug

find out more
Animals
Jellyfishes and corals

Snakes

Snakes are reptiles that have no legs. A number of snakes are burrowers, but most live on the surface of the ground. A few climb into trees, or even live in the sea.

Most snakes move by throwing their bodies into curves that push against any unevenness in the ground. Such snakes are helpless on smooth surfaces. Snakes are not slimy creatures. Their skin is dry and scaly.

Every now and then, they 'slough' (shed) their outer skin.

All snakes are flesh-eaters. They track their prey using their long, forked tongues, which flicker over the ground. The tongue picks up tiny traces of scent left by animals such as mice. Some snakes, like grass snakes, catch their food by grabbing and swallowing it. Others, like pythons, which are called constrictors, loop their powerful bodies round their prey and suffocate it. About a

third of all kinds of snakes use poison to paralyse their prey.

Snakes swallow their food whole, and have a distinct bulge in their bodies after eating a meal. To take such a mouthful they have to unhinge their lower jaws at the centre and the sides. They can move each part of the jaw independently and so 'walk' their prey into their throats.

• In cool parts of the world snakes hibernate (go into a deep sleep) during the winter. In spring they look for mates. Males often compete with each other or put on displays for females. After mating, most female snakes lay eggs, although some give birth to live young.

find out more
Reptiles

▶ Vipers are examples of front-fanged snakes. They use the fangs to inject poison as they strike their prey. Each fang is on a movable bone, which allows it to be folded away: otherwise the snake could not close its mouth.

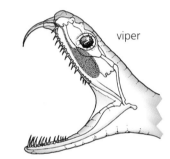
viper

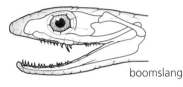
boomslang

▲ Boomslangs are examples of back-fanged snakes: they have small fangs towards the back of their mouth. This means that any prey they catch is partially swallowed before being injected with poison by the fangs.

▼ There are over 50 kinds of sea snake. Some are very poisonous, but they will only bite if they feel threatened or trapped.

sea snake

Soil

Soil is a combination of rock particles of various sizes and decayed plant and animal matter. Different kinds of rock and different climates produce different types of soils. One square metre of fertile soil can contain over 1000 million living things, yet many are too tiny to see with the naked eye.

Soil is formed when rocks are slowly broken down by the actions of wind, rain and other weather changes. Plants take root among the rock particles. The roots help to bind the particles together, and protect them from rain and wind. When plants and animals living in the soil die, fungi and bacteria break down their remains to produce a dark sticky substance called *humus*. The humus sticks the rock particles together and absorbs water. You can find out about different types of soil by looking at a *soil profile* (a sample taken from the surface down through the soil). Each profile is divided into a series of layers called *horizons*.

The rock particles in soil have air spaces between them. The larger the particles, the bigger

the air spaces between them and the faster water drains out of the soil. The air in the soil is important for plants because their roots need oxygen to breathe. The decaying plant and animal remains release minerals which are then absorbed by plant roots.

HORIZONS

O

A

B

C

R

◀ A soil profile.

• In some parts of the world wind-blown dust accumulates to form a soil called *loess*. In parts of China the loess is as much as 300 m thick.

find out more
Forests
Grasslands

The *O horizon*, the surface layer, contains many animals and plant roots. It is rich in dark-coloured humus. It is thicker in rich soils than in poor soils.
The *A horizon* still has a lot of humus, but is a paler, greyish colour because many of the minerals have been washed out by rainwater. This process is called leaching.
The *B horizon* contains much less humus, but has some of the minerals washed out of the A horizon. Any iron left here may oxidize, producing a yellow or reddish-brown colour.
The *C horizon* is where weathering takes place, and the parent rock is breaking down.
The *R horizon* is the parent (original) rock.

Songbirds

Songbirds make up the largest group of birds. Of the 8700 different kinds of bird, about half are songbirds. There is great variety within the group, from tiny wrens to large crows, and from colourful birds of paradise to the dull grey dippers of North America.

find out more
Animal behaviour
Birds
Eggs
Migration

● Not only do different kinds of songbird have different songs, but different individuals of the same kind also have slightly different songs. This helps scientists to identify an individual bird and may allow one bird to recognize another.

▶ Songbirds are found all over the world and include the majority of land birds.

Most songbirds have a well developed voice-producing organ, called a syrinx. However, although they were named for their ability to produce vocal sounds, or 'songs', not all songbirds sing, and of those that do, some, such as the common crow, make a very unpleasant noise. At the other extreme are birds such as nightingales and larks, whose songs are famed for their beauty.

All songbirds have four toes on each foot, three pointing forwards and one backwards, and are nearly all land birds. A few are found around fresh water, but most only cross the sea during migrations.

A songbird's bill is adapted to suit its diet. The bill of a hawfinch, for example, is strong enough to crack open cherry stones. Crossbills have crossed tips to the bill to help them extract seeds from tough pine cones, and tree-creepers have fine bills for probing crevices in tree bark.

Courtship and mating

In most songbirds, only the male sings a true song, although the female can produce a variety of call notes and other sounds. Along with singing, the male may strike poses and dance or move objects around, in order to attract a mate. Birds of paradise are well known for their elaborate courtship behaviour. They hang upside-down to display their beautiful feathers and sing unusual songs, which may be soft or loud. Some can sound 'mechanical' and others explosive, rather like gun-fire.

Nests of songbirds are usually quite elaborate. Most build open, cup-shaped nests. Some kinds, such as the ovenbird of South America, construct ball-shaped, closed nests, with a small entrance.

When they hatch, the young of all songbirds are naked, blind and helpless. They are usually reared by both parents. A few songbirds, such as the cuckoo finch of Africa, are parasites, laying their eggs in the nests of other birds.

black-headed weaver and nest (Africa)

waxwing eating berries (Europe and North America)

ovenbird and nest (South America)

scarlet tanager (North America)

Gouldian finch (Australia)

song thrush breaking snail shell open (Europe)

Spiders and scorpions

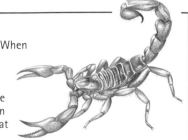

▶ A brown scorpion. When scorpions mate, they perform a special 'dance'. In the dance the male positions the female over the sperm he has dropped so that she can collect it.

Spiders and scorpions are related to each other. Both have the segments of their bodies grouped into two main parts, and have eight legs. Both are also hunters, feeding on small prey, which they first poison and then eat.

• Over 30,000 kinds of spider are known; more are discovered every year. Scientists have discovered fewer than 1000 kinds of scorpion.

garden spider

Many people are frightened of spiders and scorpions, but although there are a few aggressive kinds, most are harmless to humans.

Spiders are found in most parts of the world. All spiders can make several kinds of silk, which they use in a variety of ways, including making traps, binding their prey, and protecting their eggs. Spiders have many ways of catching a meal. Apart from the familiar webs that many spiders use to catch their prey, some spiders hunt on the ground, some leap after their prey, and some spit silk to entangle it. Others drop a net-like web onto their meal, while bolas spiders swing a sticky 'fishing' line in order to trap it. Mating can be dangerous for a male spider, which is almost always smaller than his mate. The female often eats the male after mating but nevertheless she takes good care of their eggs and young.

Scorpions are found in warm parts of the world. They hide during the daytime, but at night they come out to hunt. They feed mainly on large insects, which they kill quickly using the poisonous sting at the tip of their tail. Eating the meal usually takes them several hours. Female scorpions do not lay eggs, but produce living young, which they carry about on their back for about two weeks.

find out more
Animal behaviour
Animals

▶ Tarantulas live in the south-west of the USA, in Mexico and in South America. They hunt mainly at night, catching insects and sometimes small frogs, toads and mice. They are not dangerous to humans.

Sponges

Sponges are a very simple kind of animal. They live in the sea or freshwater, and attach themselves to rocks or other parts of the sea-bed. Sponges range in size from 1 centimetre to 1 metre or more, and come in many different colours and shapes. Some are flat, some are like vases or chimneys, and some have branches like trees.

Sponges are made up of individual cells, each of which can survive on its own. If a sponge is broken up, the cells will soon come together to make a new sponge. Each group of cells in a sponge has a different job to do. These different groups can work together, even though sponges do not have a brain or nervous system.

Although sponges look more like plants, they cannot make their own food as plants do. Like other animals, they need to eat things that have once been alive. Sponges feed on tiny food particles in the water. They also reproduce sexually in much the same way as other animals. The larvae (young) of sponges can swim through the water, although adult sponges do not move.

▲ Organ-pipe sponges growing on coral.

• There are about 10,000 different kinds of sponge.

• Bath sponges are actually the skeletons of certain kinds of sponge. In these sponges the skeleton is made of a springy material called *spongin*. Most other sponges have hard skeletons.

find out more
Animals
Cells

Squids and octopuses

Squids and octopuses, along with their relatives cuttlefishes, belong to a group of molluscs called cephalopods. Unlike many other molluscs, such as snails and mussels, cephalopods do not have an outer shell to protect their soft bodies.

All cephalopods can escape their enemies quickly using jet-propulsion: they squeeze water out rapidly through the end of their bag-like bodies, and the force of this moves them forward.

Squids live in the open seas whereas octopuses are found mostly in shallow waters, hiding in rocky dens. Cuttlefishes also live in shallow waters, and hide by burying themselves in sand.

Cephalopods are hunters. Octopuses catch their prey with their eight arms, which are covered in suckers. Squids and cuttlefishes also have eight short arms, but they use two long tentacles to catch their prey. All cephalopods bite their prey, injecting it with a poison that makes it unable to move.

With their large, complex brains and big, sensitive eyes, cephalopods are able to spot danger quickly. They can also camouflage themselves. In their skin are bags of pigment that can swell and shrink in less than a second. This rapidly changes their body colour, warning off enemies or disguising the cephalopod in the flickering light of the ocean. If an enemy does come close, they can instantly release ink, creating a 'smoke screen' of black ink.

squid

octopus

• There are about 350 different kinds of squid, about 150 different kinds of octopus, and about 14 kinds of cuttlefish. The largest cephalopod is the giant squid, which can grow to an overall length of about 20 m.

find out more
Animals
Oceans and seas
Prehistoric life

Starfishes and sea urchins

Starfishes are not really fish, but members of a big group of animals called echinoderms, which means 'spiny skins'.

starfish

Starfishes and their relatives are not obviously alike, but they have one feature in common: radial symmetry. This means their body is based around a circular plan, and if you divide it in half from above, the two halves are more or less identical.

Most starfishes have five arms. A few have a 'leading arm', but most move in any direction. If a starfish loses an arm, it can grow a new one. On the underside of starfishes and over the bodies of their relatives there are hundreds of little, soft bumps called tube feet. Each one is like a balloon, filled with sea water. Most have a sucker on the end, so the creature can hold food or grip onto the sea bed to walk.

Sea urchins are more rounded than starfishes, but you can imagine them as starfish with their arms tucked underneath. The structure of the arms can be seen on the shell as five double rows of tiny holes. Their tube feet emerge from the holes.

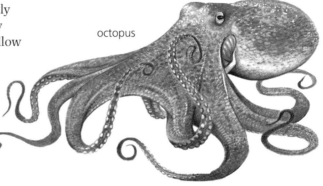

▲ Sea cucumbers are related to starfishes. Like starfishes, most kinds have five rows of tube feet, which they use to creep about on the sea bed.

► Sea urchins clearly belong to the echinoderm ('spiny skin') group of animals. They are sometimes protected by poison as well as spines. Swimmers must take care not to brush against them, nor to tread on them, because it can be very painful to have a spine in your foot.

• There are about 1600 different kinds of starfish in the world.

find out more
Animals
Prehistoric life

Swimming and diving

Swimming and diving are ways in which an animal pushes itself through water. A wide variety of animals can swim, including hedgehogs, cows and even moles. Many animals dive to find food or safety.

In order to swim, an animal must push the water back (to move forward) and down (to stay afloat). To thrust against the water effectively, it needs to maintain a good streamlined shape (more or less torpedo-shaped) so the water flows smoothly round its body.

Moving through water

Often an animal's feet, or sometimes the whole of its legs or arms, are shaped like paddles to provide a greater surface area to push against the water. Turtles and seals have flippers; dolphins and whales use their large tail flukes. Ducks, seagulls and otters have webbed feet. Many small shrimps and water insects have fringes of stiff bristles on their legs which produce a similar 'paddle' effect.

Although a few kinds of fish use their fins as paddles, most use them mainly for balancing and braking. To push themselves along most fishes move their bodies and tails from side to side.

Humans have developed several different techniques of successful swimming. The commonest of these are breast-stroke, front crawl and backstroke.

A few swimmers, such as jellyfishes and squids, use water-jet propulsion. This method of

• The fastest fish is the sailfish, which can swim at speeds up to 110 km/h.

• Sperm whales are the deepest-diving mammals. They often dive to 360 m and have been known to reach depths of over 1000 m. They normally remain underwater for between 20 and 60 minutes.

▶ About 80 kinds of bird normally dive for their food. The common kingfisher, shown here, plunges through the water's surface at great speed to catch fish and water-living insects.

swimming involves taking water into their bag-like body and then forcing it out again, which propels them forward.

Diving

Many water-living animals and some kinds of bird dive below the surface waters to find food. Animals with gills have no problem when they change level because they get their oxygen from water. But mammals, which get their oxygen from air, have to protect themselves against a dangerous build-up of nitrogen gas in the blood. To do this they breathe heavily before they dive. This transfers oxygen to their muscles, where it will be needed, and reduces air in the lungs. With less air in their lungs they also lose some of their buoyancy (ability to float), so diving becomes easier. All deep-diving mammals are able to hold their breath for a long time. They regain the lost oxygen by panting as soon as they reach the surface and can breathe again. Humans can only stay underwater for as long as they can hold their breath – about three minutes. However, they can stay longer if they breathe through a snorkel or aqualung.

Floating

Many fishes are able to float in the water with the aid of a *swim bladder*, a gas-filled bag inside their bodies. Sharks, however, have no swim bladder, and have to keep swimming to stop themselves from sinking.

◀ Many kinds of fish live in groups called *schools* with others of the same age and species (kind). Fishes in a school behave as a single unit. When threatened by danger, they swim close together, twisting and turning like a single fish.

Tapirs

Tapirs are pig-sized, hoofed mammals that live in tropical forests. Three kinds of tapir live in Central and South America, while a fourth lives in South-east Asia. All tapirs have a flexible snout, like a short trunk. This snout is made up of the nose and the upper lip. Tapirs use their snouts to grasp hold of vegetation.

Scientists think that tapirs probably look like the ancestors of horses and rhinoceroses, to which they are related. Tapirs probably have not changed much for millions of years. They are among the most primitive of the large mammals. They are shy animals, and usually live alone. They are active at night.

Tapirs have heavy, short-legged bodies, which help them to push through the dense jungles where they live. They never go far from water, and if danger threatens they dive in and can stay underwater for some time. The main natural enemies of the tapir are the jaguar in Central and South America, and the tiger in Asia. Tapirs are also hunted by humans, but the greatest threat to their survival comes from the destruction of the forests where they live.

▶ The Malaysian tapir is the largest tapir. It can grow up to 2.5 m long. It is also the most striking looking of the tapirs. The others are all a plain grey or brown colour. However, all tapir young have spotted and striped coats.

find out more
Forests
Mammals

Teeth

find out more
Skeletons

• If you do not take proper care of your teeth, they will decay. The best way to prevent tooth decay is to avoid eating too much sugar and to eat plenty of raw fruit and vegetables. You should also brush your teeth regularly.

Your teeth are the hardest part of your body. You use them every day for holding, cutting and chewing food, so they have to be very strong.

Most vertebrates (animals with backbones) have teeth. Different animals' teeth are adapted to different jobs.

Our upper and lower front teeth have straight chisel-shaped ends. They are called *incisors* and are used for slicing mouth-sized pieces from food. On each side of the incisors is a single pointed tooth. This is a *canine* tooth. The canines are more developed in meat-eating animals and are used for puncturing the skin of prey.

Our back teeth, behind the canines, are broad and bumpy. These are called *molars*. When you chew, you rub the upper and lower molars together, grinding food into small pieces for swallowing. In plant-eating animals the molars are large and have many ridges to enable them to grind tough grass and leaf material. Flesh-eating animals have special molars for chewing meat with the sides of their jaws. The edges of these teeth slide past each other like the blades of shears.

The part of a tooth that you can see is called the *crown*. Below the crown, hidden in the gums, is the tooth's root. This fixes the tooth firmly into the jawbone. Most of the tooth is made of hard material called *dentine*. The crown is coated with an even harder material called *enamel*. Inside the tooth is a cavity filled with pulp. The pulp consists of blood vessels and nerve fibres.

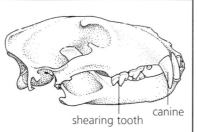

▲ Lion (carnivore)
Carnivores have long canines to kill prey and hold onto them. Massive shearing teeth crack bones and cut flesh.

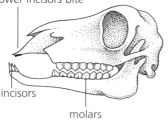

▲ Sheep (herbivore)
Herbivores cut grass by moving their chisel-edged lower incisors sideways across a thick pad on the upper jaw.

Trees and shrubs

Trees and shrubs are woody-stemmed plants. Trees usually have a single strong, woody trunk, supporting a mass of branches. By contrast shrubs often have a number of stems, giving them a bushy appearance.

Trees are the largest land plants. By growing tall, they can reach above other plants to the sunlight they need for growth. Their height also helps protect their leaves from ground-living browsers (animals that eat leaves and grass).

Shrubs or trees?

There is no real difference between shrubs and trees. The term 'shrub' is generally used for plants under 7.5 metres tall; anything taller is called a tree. Some species, such as hawthorn and hazel, can stay at shrub size if they are cut regularly or browsed by animals, but if they are left to grow they will eventually reach tree height. High in the mountains or on exposed coasts, even pine or oak trees will grow in a low, shrubby form.

Conifers and broadleaves

Trees and shrubs belong to two main groups of plants. Conifer trees, which have cones and needle-like leaves, belong to a division of the plant kingdom called the cone-bearing plants or gymnosperms. There are about 500 species, including the world's tallest trees (coastal redwoods) and the most massive ones (giant sequoias). Conifers are only distantly related to trees belonging to the flowering plants or angiosperms. These generally have wide, flat leaf-blades, and so are called broadleaves. Common broadleaf species include temperate trees such as the oak, ash, sycamore and holly. Palms, aloes and yuccas form a separate group of angiosperms.

   Trees that shed all their leaves for part of the year (generally the winter, but it can be the dry season in hot countries) are said to be *deciduous*. Shedding leaves helps protect the tree from frost damage or from drought. Other trees are *evergreen*, keeping some leaves on their branches throughout the year. Most deciduous trees are broadleaves, but a few conifers, such as larches, shed their leaves in winter.

Structure

Trees and shrubs have essentially the same structure, the main difference being the size of

the stem or trunk. The wood that makes up the stem, branches and parts of the roots is made from a mass of tube-shaped cells that transport water and food up and down the stem. Some of these tubes become thickened with a material called *lignin*. It is lignin that gives wood strength.

Roots

Usually there is as much mass of tree beneath the soil as above ground. Roots spread through the soil like underground branches. They have two basic functions. One is to form an anchor to stop the tree being blown over. The other is to absorb water and minerals from the soil. Most trees have a fungus partner that lives in their roots and helps them take the goodness from the soil.

Stem

Most of the stem of a tree or shrub is made of dead wood, which supports the plant even in strong winds. However, immediately beneath the bark is a zone of living cells. A band of tubular cells in this zone transports water and mineral salts up and down the height of the tree.

   The tubes are of two types. *Xylem* tubes carry water and minerals up the tree from its roots. The old, dead xylem tubes form the *heartwood* supporting the tree. Sugars made in the leaves are transported down the tree in another set of tubes called the *phloem*. The sugars provide energy where the tree needs it, or can be stored as starch in the trunk or roots. Each spring, the tree produces new sapwood to carry sap to the opening buds.

▲ Broadleaved trees like oak (above), beech and ash are quite slow-growing. They may take 150 years to reach their maximum height.

▼ Most conifers keep their leaves all year. A conifer's Christmas-tree shape helps it to shed snow, and its tough leaves can survive cold and ice.

Bark is the rough, grooved layer of dead, corky wood which covers the surface of trees and shrubs, protecting the living wood beneath. New bark is produced by a ring of living cells called the *bark cambium*. Older bark dies as it is pushed outwards by the expanding trunk.

Leaves

Leaves are tiny chemical factories, which make food for the tree in a process called *photosynthesis*. They take in carbon dioxide from the air and water from the soil, then combine them to make sugar. Light energy from the Sun powers this process, so the leaves are arranged to receive as much light as possible.

Leaves continuously leak water into the air by a process called *transpiration*. This sets up a suction, which draws water up from the roots through the living xylem tubes, like sucking water up through a straw. In the dead xylem, the cells gradually become blocked. This gives the trunk strength.

Fruits and flowers

Trees and shrubs reproduce (multiply and spread) by seeds. Conifers contain their seeds in a cone. When ripe, the seeds are shaken out and spread by the wind.

Like other flowering plants, broadleaves produce flowers. These may be colourful blossoms, or wind-pollinated flowers such as catkins. Some trees and shrubs, for example oak and hazel, produce separate male and female flowers. After flowering, the trees produce fruits, which contain the seeds. The fruits can be berries, nuts or dry, papery fruits which are spread by the wind.

Some shrubs or trees also spread using creeping underground stems which give off new trunks at regular intervals. All the elm trees in a hedgerow, for example, may have spread from a single tree.

The value of shrubs and trees

Trees and shrubs are valued by humans for their wood, and their fruits and leaves provide food for people and for many animals. When growing as forests, trees form an entire ecosystem, providing food high above the ground and creating a damp, shady woodland floor where other plants and animals can live.

People grow shrubs and trees in a tight mass to produce hedges. These act as a living green fence to control farm animals. Hawthorn is the commonest hedging shrub in Europe, sometimes with blackthorn or holly. Because they are expensive to look after and because farmers want larger fields, hedges are rapidly disappearing. More than a tenth of all English hedges was lost in the 1980s.

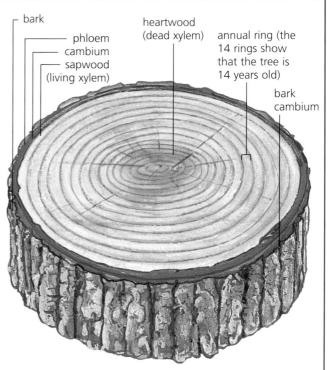

bark
phloem
cambium
sapwood
(living xylem)
heartwood
(dead xylem)
annual ring (the 14 rings show that the tree is 14 years old)
bark cambium

▲ Cross-section through a tree trunk. A ring of living cells outside the sapwood called the *vascular cambium* produces new xylem towards the inside of the trunk and new phloem on the outside. The xylem tubes grow to different sizes at different times of the year. This shows up as a series of rings through the wood. By counting these 'annual rings', it is possible to tell the age of a tree.

Cork

The light, waterproof layer of tree bark is made of cork. The cork we use to make bottle stoppers and tiles comes from the cork oak tree, which grows in the Mediterranean region. It has a thick, soft bark, made mostly of cork.

find out more
Flowering plants
Forests
Fruit
Plants
Wetlands

▼ A hedgerow in Clwyd, North Wales. Hedgerows are very important for wildlife. A good hedgerow is an excellent habitat for wild flowers, insects, birds and small mammals.

Tundra

find out more
Deer
Dogs
Ducks, geese and swans
Mice, squirrels and
    other rodents
Wading birds

Tundra is the name given to Arctic areas that are free from snow and ice for only a few months of the year. The word tundra means 'treeless', which describes the area rather well.

◀ The tundra in summer, Alaska, USA. Although the summer is short, the Sun shines for 24 hours a day. The lush green vegetation provides plentiful food for the caribou (reindeer) that travel north each summer to breed.

Tundra forms a broad belt running through the far north of Europe, Asia and America, just south of the Arctic ice sheet. Tundra also occurs on the edge of Antarctica, and in high mountains beneath the height at which snow lies all year round.

Animals and plants

Lichens, mosses, grasses and sedges are the main plants of the tundra. But there are also a few flowering plants and small shrubs that can withstand the winter cold and flower in the short summer. Animals that live in the tundra all year include voles, shrews and lemmings. In winter they tunnel beneath the snow and dig up plant roots for food. Arctic foxes also stay all year, but polar bears wander more widely, hunting seals on the ice sheets. Musk oxen are pony-sized animals, related to sheep. They are found in Canada and Greenland. They have long woolly coats and massive horns. They protect themselves from wolves by bunching themselves together in a circle, horns outwards.

When summer arrives, animals from further south come to the tundra to benefit from the rich feeding that is briefly available. Huge herds of reindeer (caribou) wander north, followed by packs of hungry wolves. And birds such as willow grouse, geese, swans, plovers and other shorebirds fly north to breed.

Turtles and tortoises

- Turtles live in the sea, tortoises live on land, and terrapins live in fresh water. There are about 240 different kinds altogether.

- Large land tortoises live for a very long time – some may live for over 200 years.

- Turtles range in size from 8 cm to over 2 m. The biggest leatherback turtles may weigh up to 590 kg.

▼ A red-eared terrapin.

Turtles and their relatives the tortoises and terrapins are among the most ancient of reptiles. They have survived since the days of the dinosaurs and have changed very little since then.

Turtles, tortoises and terrapins are easily recognized because of their armour-like shell. The shell encases and protects their body, and only their head, tail and legs show on the outside. Land tortoises can pull all of themselves into their shell. The shell is made of bone protected on the outside with horn. It is in two main pieces, one part covering the back, the other protecting the underside.

Tortoises, turtles and terrapins do not have any teeth. Instead they have sharp-edged, horny jaws with which they cut and tear their food. Tortoises are mainly plant-eaters, while terrapins feed chiefly on the flesh of various small creatures. Sea turtles eat sea grasses, seaweed and some animals.

After mating, females dig a hole in sand or soil with their hind legs, and then lay their eggs. Some kinds lay very few eggs while others lay up to 200 in a single clutch. Once the eggs are laid, the female covers the nest and leaves it. The young

◀ Hawksbill turtles are quite rare today because their shells have been used for making and decorating things. Use of 'tortoise-shell' is now illegal in many countries.

usually take several months to hatch out of their shells. They then have to fend for themselves.

▼ Hermann's tortoises are protected today because so many have been taken as part of the pet trade.

find out more
Reptiles

Viruses

Viruses cause many diseases in animals and plants. They are much smaller than bacteria, and cannot even be seen through an ordinary microscope. In fact, no one saw a virus until the electron microscope was invented in the 1930s.

Viruses are made up of a coating of protein around the material needed for reproduction (genetic material). They can only reproduce by entering the nucleus of a living cell and inserting their genetic material into the genetic material of the cell. The cell then produces copies of the virus. With some kinds of virus the cell then bursts, releasing the copies of the original virus. These viruses then go on to infect other cells.

Viruses come in a variety of shapes and sizes. Some are round, some are shaped like rods, while others have more complicated shapes. Some are so small that 300 million of them could fit onto one full stop. Even the largest are smaller than one thousandth of a millimetre in length.

Unlike bacteria, viruses cannot be killed by antibiotics.

But it is possible to be vaccinated against many of the diseases that are caused by viruses.

▼ A group of viruses, magnified many thousands of times. This image has been coloured to make the viruses easy to see. Their blue centres are the genetic material. The viruses shown here are adenoviruses, which can cause cold-like symptoms such as a runny nose.

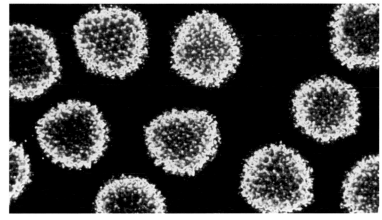

Some human diseases caused by viruses
AIDS
Cancer (some forms)
Chickenpox
Common cold
Glandular fever
Influenza (flu)
Measles
Mumps
Polio
Rabies

find out more
Bacteria
Cells
Immunity

Wading birds

Wading birds are found along coasts worldwide, and on shallow inland waters, estuaries, marshes and mud-flats. Where food is plentiful, their numbers may build up to thousands. Long legs enable them to wade into water, and long bills help them to pick up small creatures, even those that are hiding in the mud.

Most of Britain's 'waders' are sandpipers and their relatives. In North America these are called 'shorebirds', while the word 'waders' is used for herons and egrets.

The size and feeding methods of wading birds vary. Snipe have extremely long, straight bills and probe deep into the mud for worms. Oystercatchers have long, strong bills for prising open mussels, clams and other shellfish. Plovers have short, stubby bills. They run, stop, bend and seize their prey from the surface.

Herons are large with long necks, legs and toes for climbing in trees and among waterside plants. They eat animals, often fish, which they grasp in their dagger-like bills.

Many wading birds live in the tundra of the far north where food is plentiful in summer. The summer is short, but there is very little darkness and the young grow fast. Before the Arctic winter returns, the birds fly south.

◀ Wading birds find food in many different ways. Curlews probe deep into the mud (1). Pratincoles feed in the air (2). Turnstones search under stones (3). Phalaropes feed from the surface water (4), and avocets filter food from water (5).

find out more
Birds
Coasts
Migration
Seashore
Tundra

Vultures *see* Hunting birds • **Wallabies** *see* Marsupials • **Walruses** *see* Seals • **Wasps** *see* Bees, etc.

Weasels and their relatives

Although at first glance they look very different from each other, weasels, badgers, otters and skunks all belong to the weasel family.

- The weasel family contains 64 members, including 9 types of badger, 13 otters and 9 skunks. The rest of the family is made up of a variety of animals, such as wolverines, weasels and many weasel-like animals, including stoats, polecats, martens, ferrets and minks.

▶ Weasels are very fierce hunters. Some of them are so small that they are able to chase mice and voles down their tunnels.

Many of these animals, including some weasels, are small, active creatures with long bodies and short legs. But others, such as wolverines and badgers, are larger and stockier. Members of the weasel family are found in most parts of the world and in many different habitats, both on land and in water.

Weasels

Weasels are fierce hunters, feeding mainly on small rodents. Their prey is killed with a bite to the base of the skull. They are often active during the daytime, for though they hunt mainly by scent, they have good senses of hearing and sight as well.

Young weasels may remain with their mother after they are weaned, and can sometimes be seen hunting together.

Skunks

Skunks live in the Americas. Most kinds are active at night, when they hunt for small rodents, insects, eggs, birds and plants. Their fur is black with white stripes or spots; this colouring acts as a warning to enemies. If a skunk is attacked, it defends itself by banging its front feet loudly on the ground and then doing a handstand and squirting a foul-smelling fluid from stink glands just beneath its tail. Most animals leave skunks well alone.

find out more
Mammals

▶ The striped skunk is the most common kind of skunk in the USA. Its stripes act as a warning to enemies.

Otters

Otters are more at home in the water than on land. Kept warm and dry by their dense waterproof coats, otters can close their nostrils and remain

▶ When it is swimming, an otter's long whiskers can feel movements in the water. This enables it to hunt in murky streams or in the dark.

underwater for up to 6 minutes. An otter's streamlined shape enables it to swim at speeds of up to 12 kilometres per hour.

River otters hunt small fishes, frogs and other water animals, which they catch in their mouth. Clawless otters, which live in tropical Asia and in Africa, use their sensitive fingers to feel in mud or under stones for crabs and similar animals.

Sea otters also use their forefeet to capture prey such as sea urchins, molluscs and other shelled creatures. The otter breaks open the shells by banging them on a stone which it rests on its chest.

Badgers

Badgers are shy creatures and are active at night. They are good diggers, using the long claws on their forelimbs to dig out food or make burrows called sets. European badgers usually live in groups of up to 12 animals. A set may be occupied for many generations and contain over 100 metres of tunnels, as well as a large number of entrance holes and living and sleeping chambers.

In spite of their powerful teeth, badgers feed mostly on small prey and plants. Their eyesight is poor, but they have excellent senses of hearing and smell. Badgers have special glands under their tails with which they mark or 'musk' familiar objects. Some kinds of badger have an unpleasant-smelling musk which they use to defend themselves.

▲ Despite having powerful teeth and claws, badgers are generally peaceable animals.

Wetlands

Wetlands are damp, boggy areas where water lies on the surface, forming lakes or pools, or where plants have grown out into open water to form marshes or swamps. Wetlands support huge numbers of plants and animals, and their value to humans is enormous. For example, two-thirds of the fish caught around the world began their life in wetlands.

Wetlands are always changing. Once wetland plants begin to grow, they gradually build up and stabilize the ground, until eventually plants that grow on dry land can move in. In the same way, salt-marshes along coasts gradually extend out to sea. Wetlands disappear naturally as they become silted up, but at the same time new wetlands form in other places.

Types of wetland

There are many different kinds of wetland. Some are formed naturally, while others result from human activities.

Where rivers meander slowly over large flat flood plains, the slow-moving water drops the fine particles of soil and rock (sediment) into the water, and mudbanks gradually build up. River bends may get separated off, and form marshy pools or lakes. Mud also builds up on the shores of estuaries, where rivers meet the sea. At the mouth of the largest rivers, this mud forms vast fan-shaped deltas. Plants which can cope with occasional flooding by salty sea water grow here, forming *salt-marshes*, which are home to many wading birds and other animals.

The coasts of some tropical seas are fringed with *mangrove swamps*. Mangroves are trees adapted to live in wet, salty places. Their roots stick up into the air from the mud. The plants take in oxygen from the air through these roots, because there is very little oxygen in the wet mud. Mangrove swamps are home to many different animals. Fiddler crabs scuttle about the mud, scavenging for the dead remains of plants and animals. Millions of tiny creatures live in the mud, which is enriched by dead mangrove leaves. The warm, shallow waters are a nursery for the young of many ocean fishes, which are hunted by crocodiles, alligators and birds such as storks, ibises and herons.

Large areas of the cooler parts of the world are covered in *peatlands*. These are formed by remarkable plants called bog mosses, which are able to grow in waterlogged areas. As they grow, bog mosses trap water amongst their leaves, like a natural sponge. In the waterlogged soil of the peatlands, plant material does not rot away. Bog mosses pile up and form a spongy material

• People are slowly learning the usefulness of wetlands. In some places, beds of reeds are being planted to act as a natural filter for sewage, instead of building expensive treatment works.

find out more
Conservation
Ecology
Moors and heaths
River life

▼ Lesser flamingos on Lake Nakuru, a salty lake in Kenya, East Africa. Millions of birds come to the lake because the shallow water is full of the tiny insects, worms and crustaceans that they feed on.

◄ An American alligator in the Everglades, a huge area of marshland in southern Florida, USA. A large part of the Everglades is now a national park. Many kilometres of wooden walkways direct people along tracks without damaging the habitat.

animals and plants. They soak up water during storms and let it drain away gradually. This reduces the effect of floods downstream. Peatlands 'lock up' carbon dioxide released by burning coal and oil, and so help reduce global warming. Mangrove swamps help to protect tropical coastlines, by forming a natural barrier against severe storms and hurricanes. They are also important to the fishing industry, as many commercially caught fish breed in them.

In some places, wetlands ensure that the water supply for local cities is good enough to drink. The mud acts like a water filter, removing impurities from the water that passes through.

Saving wetlands

In recent years, humans have been destroying wetlands at an alarming rate. It is estimated that half the wetlands that once existed have been lost, and many more are under threat. Many wetlands have been drained so that the land they occupy can be built on or farmed. Other areas have been filled in as rubbish dumps. Many wetlands have become polluted, as the rivers that feed them have picked up pesticides and other chemicals. Some peatlands are being destroyed by large peat-fired power stations, which use up the peat much faster than it can be replaced. Peat is also used by gardeners, which adds to the destruction of wild peatlands.

Conservationists and scientists are persuading governments to set aside wetlands for wildlife, because of their valuable role in protecting the environment. But simply leaving wetlands alone is not enough. Many would gradually silt up and disappear. Their water supply must be carefully controlled to prevent this, and to stop the water draining away to surrounding areas.

known as peat. Peatlands are important areas for many breeding birds and insects.

Some wetlands are made by human activity. Disused gravel pits fill with water, and marshes develop around the edges of reservoirs. Many of these make valuable nature reserves. In warm countries, flooded paddy fields, made for growing rice, are home to fishes, egrets and herons.

The importance of wetlands

Wetlands are enormously rich food sources. Many small animals live in the soft ground, and provide food for shrews, frogs, toads and long-legged shore birds such as sandpipers and curlews. Insects thrive in wetlands, attracting insect-eating warblers and flycatchers. Seed-eating birds come to wetlands for the seed heads of the reeds and rushes, and large mammals come for water. Because of the water and rich feeding available, wetlands are important stopover places for millions of migrating birds.

Wetlands are valuable for humans as well as

▶ Mudskippers are small fishes that live in mangrove swamps, scuttling over the mud on their front fins. They can stay out of water for long periods, breathing a mixture of air and water stored in their gill chambers.

Whales and dolphins

Whales may look like fishes, with smooth, streamlined bodies perfectly adapted for life in water, but in fact they are mammals – just like us. They are very intelligent animals, and they produce a single live young, which they feed on milk.

Smaller whales are generally known as dolphins and porpoises, but in most respects they are similar to their giant relatives.

Whales

There are two main types of whale. The *toothed whales* include the smaller dolphins and porpoises, as well as giants such as the sperm whale. These animals use their pointed teeth to catch and hold fish or squid. The *baleen whales* are all large and include the right whale and the enormous blue whale. These whales feed on shrimp-like creatures called krill, which they sift out of the water with huge fringed plates in their mouths. The plates are made of baleen, a material similar to our fingernails.

Whales live all their lives in water. They breathe through a blow-hole on top of their head and can only breathe when they come to the surface.

Whales can hear very well (even though they have no visible ears). Some use sounds we call whale songs to communicate with each other. They may also use ultrasound (sound waves that humans cannot hear) to locate obstacles and food.

Whales have always been an important source of food and oil to Arctic peoples, but commercial whaling has brought many kinds close to extinction. Most whales are now protected, although a few countries continue to hunt them.

Dolphins and porpoises

Dolphins are small, slender whales, some of which are capable of swimming at speeds of about 50 kilometres per hour. Most of them have a large, curved fin in the middle of the back that helps to stabilize them. They usually live in the open sea, though some come into inshore waters.

Most kinds of dolphin have a long snout crammed with up to 200 small pointed teeth, ideal for catching fish. The largest of the dolphins, the killer whale, sometimes hunts in packs to catch much bigger animals.

Porpoises differ from dolphins in a number of ways. They are generally smaller, and have rounded faces. They are all quite playful and can often be seen in small groups, called schools, cartwheeling in the water.

▼ A sperm whale (**1**), a right whale (**2**), a dolphin (**3**) and a porpoise (**4**), drawn to scale.

- Dolphins are thought to be among the most intelligent of animals. They are sometimes kept and trained to do tricks and to help underwater engineers.

- Whales are warm-blooded, like all other mammals. They are kept warm by a layer of blubber beneath the skin.

- There are 84 different kinds of whale, including 40 dolphins and 6 porpoises. The largest of these (and the largest animal in the world) is the blue whale, which grows to about 30 m in length and weighs over 100 tonnes.

find out more
Mammals
Oceans and seas
Swimming and diving

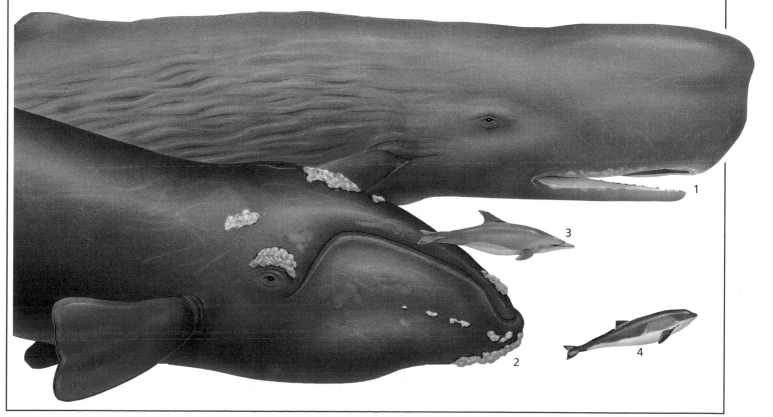

Wildlife around the world

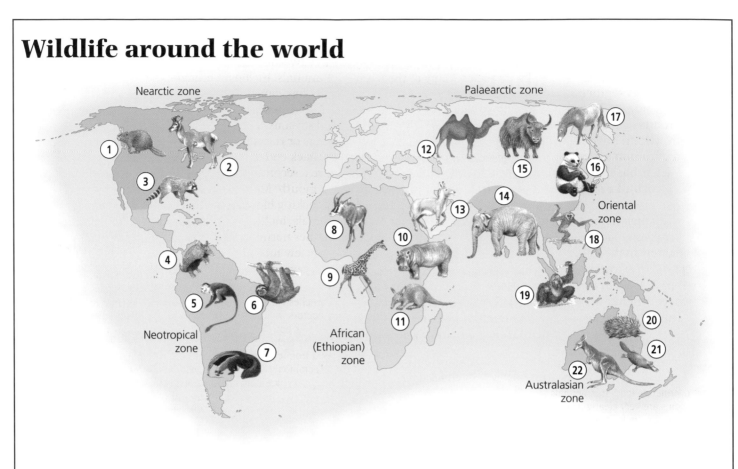

▲ A map of the world's six biogeographical zones. The animals illustrated are found only in their particular zone, and nowhere else in the world.

The word 'wildlife' means all the plants and animals in the world that are not tame or domesticated.

If you were to travel about the world, you would notice a change in the kinds of plants and animals between one area and another. This is partly because of environment and climate, and partly because the big landmasses of the continents are separated from each other by water. The plants and animals in each one have evolved (changed) over millions of years and adapted to their separate environments.

Zones of life

The world can be divided into six *biogeographical zones* ('biogeography' means the geography of living things). Each zone tends to have many plants and animals not shared by the others, although they often have related species. Even when the same kind of habitat exists in different zones, there are usually different animals of a similar kind living in it. For example, the prairies of North America and the steppes of eastern Europe and Asia are both grasslands with hot summers and cold winters. In the steppes, susliks and marmots live in basically the same way as the related prairie dogs of North America.

In a few cases the same animals may be present in different zones. This happens in northern North America, northern Asia and Europe. If you were to take a trip round these cold regions, you would find reindeer (caribou), wolverines and brown bears in the whole area, although many of the small animals are different. The reason for this is that until about 10,000 years ago there was a land bridge between north-east Asia and Alaska. Animals were able to move from one great landmass to the other. Now the sea has broken through that bridge, and the American and Asiatic animals have been separated. But they have not been apart from each other for long enough to have changed very much.

It is easiest to see the biggest differences in wildlife among the mammals. Birds, which can fly, often move between two or more zones during migration, and some kinds, such as the swallow and the peregrine falcon, are found in almost all regions. Even plants can travel between zones, as their seeds can be carried by wind, water or birds (stuck to their feet or beaks, or in their droppings).

1 mountain beaver
2 pronghorn
3 racoon
4 armadillo
5 capuchin monkey
6 tree sloth
7 giant anteater
8 roan antelope
9 giraffe
10 hippopotamus
11 aardvark
12 Bactrian camel
13 onager
14 Asian elephant
15 yak
16 giant panda
17 wild horse
18 gibbon
19 orang-utan
20 echidna (spiny anteater)
21 duck-billed platypus
22 kangaroo

find out more
Animals
Conservation
Ecology
Food chains and webs
Marsupials
Migration
Plants

Worms

Worms have long bodies, often segmented into rings, and no limbs. There are many kinds of worm, which have little in common except their basic wormlike shape.

As their name suggests, earthworms burrow in the soil. They eat soil to extract tiny scraps of food. In doing so, they turn over the soil and let air spread through it, which is very important in keeping it fertile.

▲ Earthworms are long, thin animals whose bodies are very clearly made up of a series of similar segments, or rings. The head contains a brain and a mouth, which has tiny, horny jaws.

Flatworms, which are sometimes called flukes, usually live in water. Many of them are brightly coloured and are generally less than 1 centimetre long. Some kinds of fluke are parasites on other animals, and may cause serious diseases.

Tapeworms are related to flatworms. As adults they are parasites, living and feeding in the guts of another animal. Tapeworms are so called because they are long and flat, like a tape-measure.

Roundworms are usually less than 2 millimetres long. Huge numbers of them live in soil, in the sea and in fresh water. Some roundworms, called threadworms, can be parasites of animals, including humans, and often cause diseases.

Leeches usually live in ponds and streams. They hang onto plants or rocks using a sucker at the back end of the body. Many have a second sucker around the mouth. Most leeches feed by sucking blood from other animals, including humans. Leeches normally feed once every few months, but can go for a year without feeding.

▼ Leeches were once frequently used by doctors in Europe to suck blood from sick patients. Leech saliva contains a chemical which prevents blood from clotting. Diseases commonly 'treated' with leeches were mental diseases, headaches and tumours.

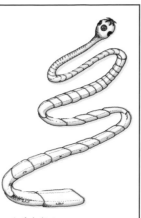

▲ Adult tapeworms cannot see, smell or hear. They are parasites and live protected inside the guts of pigs and other vertebrates (animals with backbones). Here they are surrounded by digested food, which they take in through their body wall.

find out more
Animals
Pests and parasites
Soil

Zebras

Zebras are close relatives of horses and donkeys. They are wild animals, and live on the plains and hills of eastern and southern Africa. All zebras have striking black-and-white stripes. These stripes work well as camouflage, as they help to break up the outline of the animal and so confuse hunters.

There were once four different kinds of zebra, each with a different pattern of stripes. The common zebra (also called the plains or Burchell's zebra) is still widespread, but the mountain zebra and Grevy's zebra are now rare. The quagga, which was only striped at the front of its body, was killed off by hunters at the end of the 19th century.

Zebras live in small groups of females and their young, led by a single male. The male will lead the family for about 10 years, then he will be replaced by a younger male. Sometimes zebra families form larger herds, and sometimes they join herds of antelopes and migrate long distances together to find fresh grass to eat.

Zebras use their ears to communicate. When they are angry, they flatten their ears. When they are feeling friendly, their ears stand upright.

find out more
Conservation
Grasslands
Horses
Mammals

▼ As a family of zebras moves, it keeps a strict order. A mare (adult female) leads. She is the dominant female and is followed by her young, and then by up to six more mares and their foals. Right at the back comes the stallion, the father of all the foals, who fights off predators, such as lions.

Zoos

The word 'zoo' is short for zoological gardens. Zoos are places where people can go to see wild animals from many parts of the world. The first zoos were collections of animals made by ancient kings and noblemen. Today there are many types of zoo, some specializing in particular sorts of animals.

One of the most popular kinds of zoo is the oceanarium, which keeps dolphins and other sea creatures. However, some people believe that it is wrong to keep these large sea animals captive. Another type of zoo is the safari park, in which it is the visitors who are caged in their cars, and the animals have large areas in which to roam freely.

What zoos do

Apart from keeping animals for visitors to see, modern zoos have three main functions: to study the animals and find out as much as possible about their habits; to educate visitors; and to breed from the captives for return to the wild.

Breeding is an important part of the work of zoos. Many kinds of animal have such small numbers remaining in the wild that producing young in zoos can be the only way to save them

▼ In this oceanarium, a transparent tunnel allows visitors to walk right through a large tank containing sharks, rays and other ocean creatures.

▲ An Arabian or white oryx with its calf. Arabian oryx once lived in the deserts of the Arabian peninsula, but by the early 1970s the only animals left were in zoos. They were successfully bred in captivity, and in 1982 a herd of animals was reintroduced into Oman.

from extinction. If enough animals can be bred, groups can be returned to the wild in the areas where they have become extinct. This has been done with the Hawaiian goose (nene), Père David's deer and the Arabian oryx.

▶ FLASHBACK ◀

The first zoos were collections of animals made by kings and noblemen in the Middle East and China. In the past, zoo animals generally had a short life, for the right food and conditions were seldom available. When an elephant was first brought to Britain in 1254, it was housed in a special building 6 by 12 metres, which was certainly not big enough by modern standards. For centuries most zoos were miserable prisons for many animals, which were kept on their own. Such solitary confinement is the height of cruelty for social animals such as monkeys and apes.

In recent years many zoos have realized the importance of putting the interests of the animals first. The animals are fed adequate amounts of the correct food, and their enclosures are designed to imitate as far as possible the animal's wild home. In good zoos most animals live far longer than they would in the wild. There are still some bad zoos but fewer than there used to be.

• Zoos are extremely costly places to run. The animals have to be securely housed and kept warm. Their food may also be expensive.

• The Roman emperor Nero is said to have had a tame tigress called Phoebe, while the favourite of Charlemagne, who kept many animals, was an elephant called Abul Aba.

find out more
Animals
Conservation

Index

If an index entry is printed in **bold**, it means that there is an article under that name in the A–Z section of the encyclopedia. When an entry has more than one page number, the most important one may be printed in **bold**. Page numbers in *italic* mean that there is an illustration relating to the entry on that page.

Q

quaggas *31*, 123
Quaternary *93*
quills 65, 79

R

rabbits 96
rabies 15
racoons 96, *122*
radioactive waste *91*
Rafflesia 89